ATOMIC-FORCE MICROSCOPY AND ITS APPLICATIONS

Edited by **Tomasz Tański, Marcin Staszuk** and **Bogusław Ziębowicz**

Atomic-force Microscopy and Its Applications
http://dx.doi.org/10.5772/intechopen.74139
Edited by Tomasz Tański, Marcin Staszuk and Bogusław Ziębowicz

Contributors

Milad Radiom, Gmt (Threes) Smijs, Federica Galli, Oleg Alexeevich Ageev, Marina Il'ina, Oleg Il'in, Vladimir Smirnov, Yuriy Blinov, Boris Konoplev, Tatiana Sukhanova, Arzu Erol, Tomasz Arkadiusz Tański, Boguslaw Ziebowicz, Pawel Jarka, Marcin Staszuk

First published in London, United Kingdom, 2019 by IntechOpen
IntechOpen is the global imprint of INTECHOPEN LIMITED, registered in England and Wales, registration number: 11086078, The Shard, 25th floor, 32 London Bridge Street
London, SE19SG – United Kingdom
Printed in Croatia

British Library Cataloguing-in-Publication Data
A catalogue record for this book is available from the British Library

Additional hard copies can be obtained from orders@intechopen.com

Atomic-force Microscopy and Its Applications, Edited by Tomasz Tański, Marcin Staszuk and Bogusław Ziębowicz
p. cm.
Print ISBN 978-1-78985-169-4
Online ISBN 978-1-78985-170-0

For EU product safety concerns:
IN TECH d.o.o., Prolaz Marije Krucifikse Kozulić 3, 51000 Rijeka, Croatia,
info@intechopen.com or visit our website at intechopen.com.

Meet the editors

Prof. Tomasz Tański is the Head of the Institute of Engineering Materials and Biomaterials at the Silesian University of Technology; a specialist in materials such as non-ferrous alloys, composite and nanostructured materials, manufacturing engineering; surface, properties and structures of engineering materials. He has authored or co-authored more than 300 scientific publications worldwide including 12 monographs and books, more than 70 publications in the Philadelphia list; won 18 National and International Awards and Honors; has served, or is currently serving, as a supervisor or contractor for more than 15 research and didactic projects in Poland and abroad. Prof. Tański is also a reviewer and promoter of numerous scientific papers, including 7 doctoral research papers in the field of nanotechnology and materials.

Dr. Marcin Staszuk obtained a doctoral degree from the Silesian University of Technology in 2010, in Technical Sciences in the field of Materials Engineering. He works as an Assistant Professor at the Institute of Engineering Materials and Biomedical Engineering at the Silesian University of Technology. His scientific interests include surface engineering, tool materials, light metal alloys and modern materials investigation methods. He is the author and co-author of numerous international scientific works, including 22 publications from the Philadelphia list. He has completed numerous scientific and didactic internships in Portugal, Greece and the Czech Republic. At the International Inventions and Innovation Exhibitions, he won three gold medals. He is the supervisor of numerous MA theses and the auxiliary supervisor in the PhD thesis.

Dr. Bogusław Ziębowicz obtained a doctoral degree in technical sciences in the field of Materials Science at the Silesian University of Technology. The subject of his doctoral thesis was "Structure and properties of composite materials composed of nanocrystalline powders $Fe_{73,5}Cu_1Nb_3Si_{13,5}B_9$ and polyethylene". With the aim of further scientific activity, he continued the work on dental materials - nanostructured layers on dental components. Currently, his scientific interests include primary uses of an atomic force microscope in the study of the surface of dental materials. He enjoys working with students and he is supervisor of the Student Scientific Circle Nanotechnology and functional materials. During his scientific activity, he was a participant at many scientific international conferences. He is an author and co-author of more than 60 scientific publications worldwide including 8 publications in the Philadelphia list. Privately, he loves skiing and music from vinyl records.

Contents

Preface IX

Section 1 Introduction 1

Chapter 1 **Introductory Chapter: Why Atomic Force Microscopy (AFM) is One of the Leading Methods of Surface Morphology Research of all Engineering Material Groups 3**
Tomasz Tański, Bogusław Ziębowicz, Paweł Jarka and Marcin Staszuk

Section 2 The Atomic-force Microscopy Study 9

Chapter 2 **Characterization of Single Polymer Molecules 11**
Milad Radiom

Chapter 3 **Forensic Potential of Atomic Force Microscopy with Special Focus on Age Determination of Bloodstains 27**
Threes Smijs and Federica Galli

Chapter 4 **Scanning Probe Techniques for Characterization of Vertically Aligned Carbon Nanotubes 49**
Marina V. Il'ina, Oleg I. Il'in, Vladimir A. Smirnov, Yuriy F. Blinov, Boris G. Konoplev and Oleg A. Ageev

Chapter 5 **Characterization of Multiblock (Segmented) Copolyurethane-Imides and Nanocomposites Based Thereof Using AFM, Nanotribology, and Nanoindentation Methods 69**
Tatiana Evgenievna Sukhanova, Tatyana A. Kuznetsova, Vasilina A. Lapitskaya, Tatiana I. Zubar, Sergei A. Chizhik, Milana E. Vylegzhanina, Aleksandr A. Kutin, Andrey L. Didenko and Valentin M. Svetlichnyi

Chapter 6 **High-Magnification SEM Micrograph of Siloxanes 91**
Arzu Erol

Preface

The synergy achieved in the field of new material technologies regarding the shaping of the structure and properties of the surface of materials is the result of the integration of many fields of science and technology. Many research centres in the country and in the world are continuing their efforts to thoroughly learn and describe phenomena occurring on the surface of solid bodies. Due to the development of nanotechnology in many applications such as medicine or electronics, the definition of horizontal millimetre structure, or even micrometres is not enough, because continuous technological progress creates the need to image details of smaller and more accurate dimensions. This need is met by the dynamically developing scanning tunnel microscopy (STM) and atomic force microscopy (AFM) methods.

Among many tools associated with nanotechnology, none is so well recognized and simple in terms of construction assumptions than the atomic force microscope, also called a microscope of close interactions. Despite the simplicity of the basic assumptions regarding operation, the design of this microscope is not at all trivial, and high-quality modern systems contain the latest technological solutions. Over the past two decades, the atomic force microscope has evolved from a highly specialized device into a computer-controlled microscope that can be placed on the table. During this time, the number of possible applications for this device has also increased significantly, and the development of research techniques associated with it continues. As a result, a device was created enabling surface imaging in three dimensions, but also an interactive tool that gives the opportunity to manipulate and interfere with the structure of the sample at the level of individual molecules.

This book introduces readers to the latest achievements in the field of near-interaction microscopy, presents the possibilities of using SPM microscopes both for testing and modifying the surface of materials at the atomic scale. It constitutes an interesting literature position for both research communities in the fields of chemistry, physics, biology and materials engineering.

Prof. Tomasz Tański, Dr. Bogusław Ziębowicz, and Dr. Marcin Staszuk
Division of Materials Processing Technology, Management, and
Computer Techniques in Materials Science
Institute of Engineering Materials and Biomaterials
Silesian University of Technology
Gliwice, Poland

Section 1

Introduction

Chapter 1

Introductory Chapter: Why Atomic Force Microscopy (AFM) is One of the Leading Methods of Surface Morphology Research of all Engineering Material Groups

Tomasz Tański, Bogusław Ziębowicz,
Paweł Jarka and Marcin Staszuk

Additional information is available at the end of the chapter

http://dx.doi.org/10.5772/intechopen.80446

1. Introduction

Forming part of the group of scanning probe microscopy, the atomic force microscopy (AFM) allows to imaging of surface topography of all groups of engineering materials with very high resolution, relative to all three axes. AFM enables of obtaining very large magnifications up to 10^8 and resolution of 0.1 nm in the x, y, and 0.01 nm axes in the z axis. Compared to classic scanning electron microscopy (SEM), it allows to imaging of dielectric material surfaces without the need for applying conductive layers and allows the determination of surface roughness parameters without the need for additional tests. Therefore, this method is still being developed and is recognized in many areas of science and technology. The special significance of AFM is emphasized in the research on the surface quality of functional materials, using the possibility of surface analysis at the nanometric level. The field of AFM applications, due to the enormous possibilities unavailable to other research techniques, is constantly expanded [1–4].

The idea of AFM is based on the use of interatomic short-range interactions whose intensity depends on the surface topography of the test sample subjected to the research. The analysis of the surface in terms of the shape and location of characteristic points most often carried out by means of AFM is in three modes [5–7].

It enables very accurate imaging of the surface topography of the static-contact mode (the microscopy of repulsive interactions), in which the distance between the atoms of the probe and atoms of the exanimated surface is in the range of 1 Å and the pressing force of the probe is in the range of 10-11 to 10-7 N. However, the limitation of the use of the mode in the case of materials with low hardness (thin organic layers) is the need to apply low-pressure forces by danger of damaging the sample [5].

In the dynamic-resonance mode (long-range attraction microscopy—van der Waals forces, magnetic forces, electrostatic forces), i.e., noncontact mode, the probe is located 1–10 nm from the sample surface. The mode provides lower resolution of images than in the contact mode; however, it can be used to analyze the surface of materials with very low hardness and easily deformable [6].

In intermittent-contact mode (resonant microscopy of repulsive interactions), the probe introduces oscillation at a frequency close to the resonant frequency (50–500 kHz) at a distance enabling the analysis of interactions of both long-range forces and short-range forces and the blade periodically comes into contact with the surface of the sample [7].

A common feature of all available modes is the method of image creation as a result of measuring the forces occurring between the microprobe and the surface under test. The microsphere is placed on a springy lever (beam) which is bent under the influence of interatomic forces. Deflection of the beam is processed by the photodetector for a current signal, which allows obtaining a surface image of the sample. Due to the specificity of surface imaging, AFM is effective for virtually all groups of engineering materials. This method has no limitations due to the mechanical, optical, or electrical properties of the materials [3–5, 8].

In the case of ceramic layers with high hardness, the use of AFM provides detailed information on the morphology of the surface layers obtained in PVD and CVD (chemical vapor deposition) processes produced as coatings, among others on ceramic tool materials. For example, the analysis of Ti (C, N) + Al_2O_3 + TiN coating surfaces allows to show a topography characteristic of the subsurface Al_2O_3 layer consisting of numerous polyhedra. In this case, the used atomic force microscopy enabled the imaging of the nanodimensional structures and the precise measurement of the height of the structural surface elements by profile analysis which is unachievable by the use of other surface observation techniques, e.g., scanning electron microscopy [9].

An extremely important element of a properly conducted surface topography survey is the analysis of results containing the appropriate interpretation of processes and phenomena occurring on the surface of the material being tested. AFM image analysis is carried out using programs dedicated to specific hardware applications used to analyze and measure basic parameters of the image surface, such as quantitative measurements of surface features in selected areas, statistical data such as geometric area, surface area, and the ratio of surface area to geometric surface.

The use of AFM is developed in particular in the fields of material engineering in which very precise measurement of surface quality is necessary due to both the most modern methods of obtaining surface structures and the specific use of obtained materials. For this reason, AFM is often used in the research of dental materials and, in particular, the layers of materials used in dentistry produced by the most modern methods providing very high-dimensional accuracy and surface quality. The ALD (atomic layer deposition) method plays an important role in dental prosthetics as one of the most modern coating techniques. ALD is a variant of the CVD (chemical vapor deposition) that allows the coating deposition of very accurately reflecting the topography of the substrate and ensures control of the thickness of the deposited coating with accuracy to the layer of atomic thickness, which allows coating complex-shaped surfaces [3].

In the case of layers of materials used in dentistry produced using the ALD method, the application of AFM for surface layer morphology tests is an important supplement to the surface quality tests carried out so far. It is extremely important to examine the roughness of the created layers, the low value of which makes it practically impossible to use a different test method. An important issue is also the ability to determine the quality of the layer made with very high magnification [4, 9].

The area of application of AFM is also surface investigations of materials whose physical properties (e.g., lack of electrical conductivity, low hardness) make it difficult to analyze with other methods, and determination of surface morphology is extremely important due to their potential application.

Atomic force microscope enables detailed analysis of topography of thin organic layers or composites with organic matrix and micro- and nanostructure fillers. The results of the analysis of topographic images obtained using this method allow to obtain detailed information on the state of the surface, damage on the nanoscale, and lumps of material deposited with possible agglomerates in the case of nanoparticle-strengthened composites. One of the basic advantages of AFM from the point of view of organic materials research is the possibility of imaging with a very large magnification surfaces of nonconductive materials. The complementary method of examining the surface of materials can be scanning electron microscopy (SEM) in the case of nonconductive organic materials forces coating their surface with thin conductive layers (gold, silver, and aluminum, among others) in order to perform the test [3, 4, 8, 9].

In addition, in case if thin layers with nanometric thickness (less than 100 nm) are from organic materials of mostly very low density, the embedding of the conductive layer can cause a significant decrease in the accuracy of imaging, e.g., in the form of fractures of additional conductive material. The necessity of depositing thin conductive layers on the surface of organic material also makes it impossible to carry out further research on a thin organic layer forming, for example, a heterojunction p-n due to a change in the optical properties of the system. In addition, thin layers of organic materials are easily degraded under the influence of the electron beam used in SEM. This makes the AFM an irreplaceable imaging method that provides an image of the surface of organic materials with the resolution capability of the order of a single atom. The low mechanical strength and, in particular, the low abrasion resistance, thin organic layers, and nanocomposites with an organic matrix make that in the research of this type of materials, the noncontact mode is most often used. In addition, in the case of composite layers with nanofillers, due to the high tendency to form agglomerates with sizes above 1 μm, the use of noncontact imaging avoids the risk of damage to the probe [3, 9].

An extremely important advantage of AFM is the ability to describe quantitative surface quality by determining roughness coefficients. The quantitative analysis of the surface topography of layers is carried out based on surface unevenness parameters. The basic parameters describing surface topographies are RMS (rough mean square) and Ra parameter. The RMS parameter is defined as the standard deviation from the mean value calculated from the area based on the point grid (characterized by the height of Zi) according to the formula 1:

$$RMS = \sqrt{\frac{1}{n} \cdot \sum_{i=1}^{n} (Z_i - \bar{Z})^2} \quad (1)$$

where n is the number of points, Z_i is the height of each point for the z coordinate, and $\bar{Z}$ is the average value of the sample height for the z coordinate.

AFM also allows thickness measurement of thin organic layers, which is extremely convenient especially for materials with low hardness. Methods that use the determination of layer thickness based on measurement of optical properties require complex calculations or are relatively time-consuming. Using AFM, thickness measurement is simpler and faster. The measurement consists in removing the material of the layer up to the substrate material. Then, the measurement of surface topography at the boundary of the layer and the substrate is conducted. Through the linear analysis of the height change value in the topographic profile, it allows to specify the layer thickness. However, the accuracy of the test should depend on the total exposure of the substrate and the selection of a representative place thickness test. Therefore, especially in the case of layers with complex topography, it is necessary to perform measurements several times in order to determine the statistical average [9].

Author details

Tomasz Tański*, Bogusław Ziębowicz, Paweł Jarka and Marcin Staszuk

*Address all correspondence to: tomasz.tanski@polsl.pl

Division of Materials Processing Technology, Management, and Computer Techniques in Materials Science, Institute of Engineering Materials and Biomaterials, Silesian University of Technology, Gliwice, Poland

References

[1] Reifenberger R. Fundamentals of Atomic Force Microscopy. Word Scientific; 2015

[2] Binnig G, Quate CF, Gerber CH. Atomic force microscope. Physical Review Letters. 1986; **9**(96):930-933

[3] Haustad G. Atomic Force Microscopy Understanding Basic Modes and Advanced Applications. Wiley; 2012

[4] Jand KD. Atomic force microscopy of biomaterials surfaces and interfaces. Surface Science. 2001;**3**(491):303-332

[5] Voigtjander B. Scanning Probe Microscopy. Springer; 2015

[6] Morita S, Giessibl FJ, Meyer E, Wiesendanger R. Noncontact Atomic Force Microscopy. Springer; 2015

[7] Prater CB et al. Tapping Mode Imaging: Applications and Technology. Santa Barbara: Digital Instruments; 1997

[8] Kruk T. Atomic force microscopy (AFM). The Laboratory. 2013;**18**(1):40-50

[9] Zięblowicz B, Staszuk M, Jarka P. Surface analysis using an atomic force microscope. The Laboratory. 2018;**18**(1):18-22

Section 2

The Atomic-force Microscopy Study

Chapter 2

Characterization of Single Polymer Molecules

Milad Radiom

Additional information is available at the end of the chapter

http://dx.doi.org/10.5772/intechopen.77999

Abstract

This chapter offers an overview of the use of atomic force microscopy (AFM) in polymer studies. Soft AFM cantilevers with sharp tips are useful for their relatively high spatial resolution, a few nm, and force resolution, a few tens of pN. AFM imaging is used to characterize conformational properties of single polymer chains at solid-liquid interfaces. AFM force microscopy gives molecular elasticity as well as interaction forces of single polymer chains with solids. Recent technical developments have made possible the characterization of time-resolved mechanical properties of single polymer chains, including the relaxation time and internal friction. AFM force microscopy with biomolecules, supramolecules, and mechanophores reveals the forces required for, and the kinetics of, conformational transitions and chemical reactions in these molecules at the single-chain and single bond levels.

Keywords: AFM imaging, atomic force microscopy, mechanochemistry, molecular conformations, molecular elastic response, single molecule force microscopy, single molecules

1. Introduction

From the time of its invention in 1986 [1], atomic force microscopy (AFM) has been influential in polymer studies mainly at the nanoscale. The imaging mode of AFM has been used to visualize polymer chains [2, 3], while the force microscopy mode to measure their elasticity, internal friction, and adhesion forces [4–7]. Moreover, the long-established theories of polymer mechanics and dynamics could be reevaluated and retuned to better interpret the new results obtained from AFM measurements [8, 9]. Alongside theories, computational chemistry methods have been adopted to evaluate relevant experimental parameters from *ab initio* or molecular dynamics calculations, or to model the force response of polymers with conformational transition, for example polysaccharides [10, 11].

In imaging application, polymer chains are generally adsorbed from a dilute solution. The dilute condition results in thin polymer films where the chains are isolated. The polymers are deposited on flat solids such as mica, silica (due to roughness, silica is used with thick polymers such as dendronized polymers), gold (for example, gold deposited on mica), or highly oriented pyrolytic graphite (HOPG). The individual chains are then imaged using noncontact or intermittent contact imaging modes [12–17]. Analysis of AFM images provides useful information on conformations and sizes of polymer molecules, and conformational transitions because of changing chemical environment [3, 18–21]. Examples of AFM images of double-stranded DNA [3] and four generations of a dendronized polymer [22] are shown in **Figure 1 (a)** and **(b)**. Analysis of DNA images shows the effect of chemical environment, solution as well as solid substrate, on DNA conformation and length. Processing of the AFM images of dendronized polymers show that chains thicken with generation of dendronization, while their conformations persist over longer distances.

In a seminal work, Gaub and coworkers showed that AFM can be used to manipulate proteins at single molecule level [4]. This research led to the use of AFM in polymer studies involving the extension and manipulation of single polymer chains. The measurements are realized by adsorbing a polymer film on solid from a dilute to moderately concentrated polymer solution. The tip of the AFM cantilever is then brought into contact with the solid and retracted. This process results in occasional extension of a single chain. The solid substrate and the AFM tip can be functionalized to chemically bind the polymer chains, or to tune between extension and desorption interactions [5, 6]. To model the force versus extension profiles, the polymer chain is modeled with a continuous curve, or as a series of discrete segments that are freely jointed or jointed at fixed bond angles with rotational freedom [26]. These models normally incorporate a characteristic length corresponding to entropic elasticity of the polymer and a characteristic elasticity constant corresponding to deformation of bond angles. Examples of AFM force microscopy of poly(ethylene glycol) (**PEG**) [23] and single-stranded DNA [11] are shown in **Figure 1(c)** and **(d)**. In both cases, one observes that the force increases monotonically with extension. This is because the polymer chain loses its entropy during elongation causing a restoring force on AFM cantilever. Unlike the response of single-stranded DNA, **PEG** force response shows conformational transition in electrolyte solution. The transition is absent in non-hydrogen bonding hexadecane.

Among other developments, AFM single molecule force microscopy was combined with electrochemistry to obtain sequential extension-oxidation-relaxation giving a thermodynamic cycle with a single chain of a redox polymer [27]. Using two AFMs in parallel configuration, a correlation force microscope (CFM, or correlation force spectroscopy, CFS) was developed and used to measure the dynamics of single polymer chains, namely elasticity and relaxation time [7, 28]. Furthermore, by laterally dragging single polymer chains that are covalently bound to AFM tip and adsorbed onto solid, nanoscale friction mechanisms were investigated using a single polymer chain probe [29, 30].

AFM is also used to activate chemical reactions and conformational transitions at single polymer chain level. In this case, the polymers contain force-sensitive units, which are activated by application of mechanical force. Moreover, to measure the strength of chemical bonds, one may

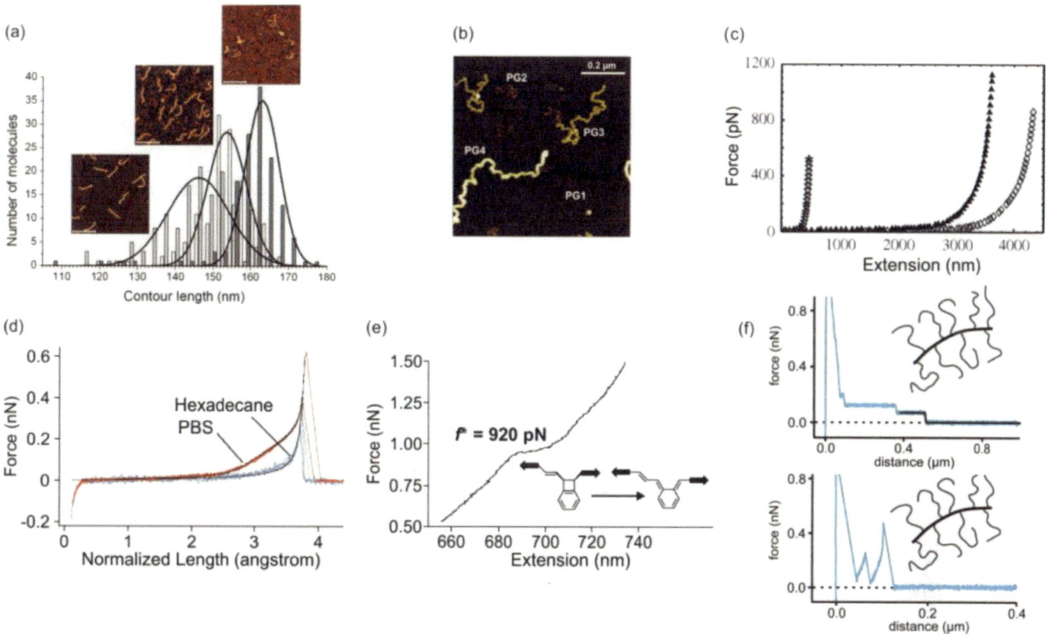

Figure 1. (a) AFM images of 500 base-pair DNA with the corresponding histograms of contour lengths. DNA deposited on (3-aminopropyl)triethoxysilane modified mica resulted in the longest length, while when deposited on mica from a solution containing Mg^{2+}, it resulted in the shortest length. Middle range length was for when DNA was deposited on mica from a solution containing Mn^{2+}. Adapted with permission from Japaridze et al. [3]. Copyright © 2016 American Chemical Society. (b) AFM image of generation 1–4 of a dendronized polymer that has two terminal amines per monomer adsorbed on mica. The image shows thickening and longer conformational persistence of the polymers with generation. Adapted with permission from Zhang et al. [22]. Copyright © 2011 American Chemical Society. (c) Force versus extension response of single-stranded DNA chains. The DNA was adsorbed on a gold-coated surface and extended in Tris buffer. Reprinted figure with permission from Hugel et al. [11]. Copyright (2005) by the American Physical Society. (d) Force versus extension response of poly(ethylene glycol) (**PEG**) polymer chains. **PEG** was deposited on a gold surface, and the force measurements were carried out in either phosphate-buffered saline (PBS) or hexadecane. The solid line shows the best fit to freely jointed chain (FJC) model, in the case of hexadecane, or two-state FJC model, in the case of PBS. Reprinted figure from Oesterhelt et al. [23] (e) Ring opening of benzocyclobutene with AFM force microscopy at a force of about 920 pN. Polymers containing benzocyclobutene units were absorbed on a silica. Measurements were performed in toluene. Adapted with permission from Wang et al. [24] Copyright © 2015 American Chemical Society. (f) Force versus extension response of poly(isoprene) with 88 kDa **PS** side chains in water and on hydrogen-terminated diamond showing steplike desorption response, and spikelike extension and detachment response. Adapted with permission from Kienle et al. [25]. Copyright © 2014 American Chemical Society.

incorporate a functional group at free end of polymer and investigate specific interactions between the group and the AFM tip or the solid. Investigation of chemical reactions at single-chain or bond level using AFM has led to insights into forces and kinetics of various chemical reactions and transitions, including complexation and coordination [31, 32], receptor-donor type interactions [33], hydrogen bonding [34], and covalent bonding [35, 36]. An example of mechanochemistry at single-chain level is shown in **Figure 1(e)**. AFM force microscopy reveals that the force of opening benzocyclobutene ring is about 1400 pN in toluene, but reduces to 920 pN with the help of an alkene lever arm in the structure of the polymer [24].

Below, I have illustrated AFM application in polymer studies with specific examples. Schematics of the AFM applications in imaging, force microscopy, and other modes are shown in **Figure 2**. The structures of some of the polymers used in the experiments are summarized in **Figure 3**. The polymers are poly(2-vinyl pyridine) (**P2VP**), poly(styrene) (**PS**), poly(ethylene) (**PE**), poly(ethylene glycol) (**PEG**), and a triblock copolymer of poly(exo-N-(2-aminoethyl)-5-norbornene-2,3-dicarboximide) and poly(exo-N-hexyl-5-norbornene-2,3-dicarboximide) (**P1**).

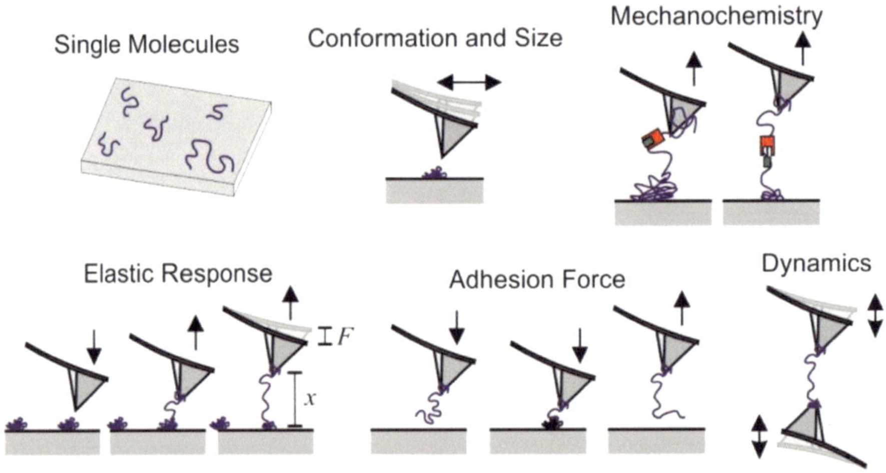

Figure 2. Schematics of AFM imaging of isolated polymer chains, mechanochemistry with AFM, AFM force microscopy of single polymer chains to obtain their elasticity or adhesion forces, and schematic of correlation force spectroscopy (CFS is a variant of AFM) to obtain dynamical mechanical properties of single polymer chains.

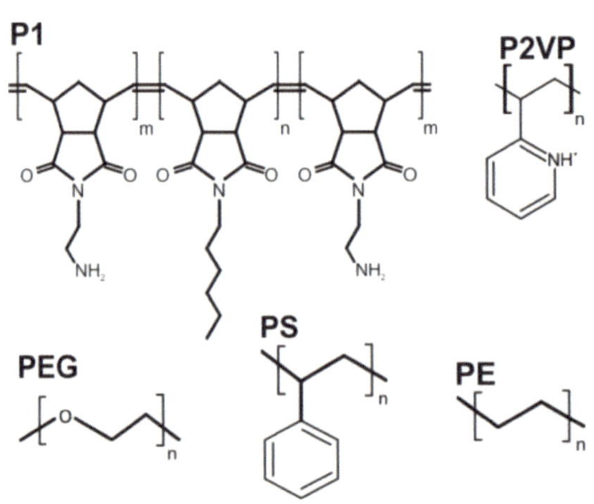

Figure 3. Chemical structure of a triblock copolymer of poly(exo-N-(2-aminoethyl)-5-norbornene-2,3-dicarboximide) and poly(exo-N-hexyl-5-norbornene-2,3-dicarboximide) (**P1**) [37], poly(2-vinyl pyridine) (**P2VP**), poly(ethylene glycol) (**PEG**), poly(styrene) (**PS**), and poly(ethylene) (**PE**). The side blocks of **P1** are about 11 monomers long ($m \approx 11$) and the middle block 544 monomers long ($n \approx 544$). The side blocks contain amine, which enhances bonding of the polymer ends to epoxy-functionalized AFM tip and solid. The covalent bonding helps pull the polymer to high forces of about 1 nN. **P2VP** is positively charged at pH 3.0.

2. Molecular conformations obtained from AFM imaging

Conformation of a single polymer chain may be interpreted in terms of average of spatial correlations between unit vectors **n** tangent to the chain. In the framework of wormlike chain (WLC) model, the average function is of the form:

$$\langle \mathbf{n}(0) \cdot \mathbf{n}(s) \rangle = \exp\left(-\frac{s}{2\ell_p}\right), \tag{1}$$

where s is the length, and ℓ_p is the characteristic decay length of the correlations, or the persistence length. Image analysis software has been developed that tracks the imaged chains and quantify their persistence lengths using Eq. (1) [38].

The correlations generally decay rapidly for thin and flexible polymers, but persist longer for thick and semiflexible polymers, such as double-stranded DNA, which have inherent bending rigidity [3, 18]. For charged polymers such as polyelectrolytes, the persistence length has a contribution from intramolecular electrostatic repulsion, which tends to expand the chain. This contribution may be controlled by pH and the ionic strength of an electrolyte solution. Odijk, Skolnik and Fixman (OSF) theory predicts that the electrostatic contribution decays rapidly with inverse of the ionic strength [39, 40]. However, experiments and simulations generally find a slower decay [18, 41, 42].

Figure 4 shows two AFM images of poly(2-vinyl pyridine) (**P2VP**) polymer chains. The dilute polymer films were prepared as follows. A solution at pH 3.0 was initially prepared by addition of HCl to deionized water. The ionic strength of this solution is approximately 1 mM. To this solution, appropriate amount of NaCl was added to set the ionic strength to 100 mM.

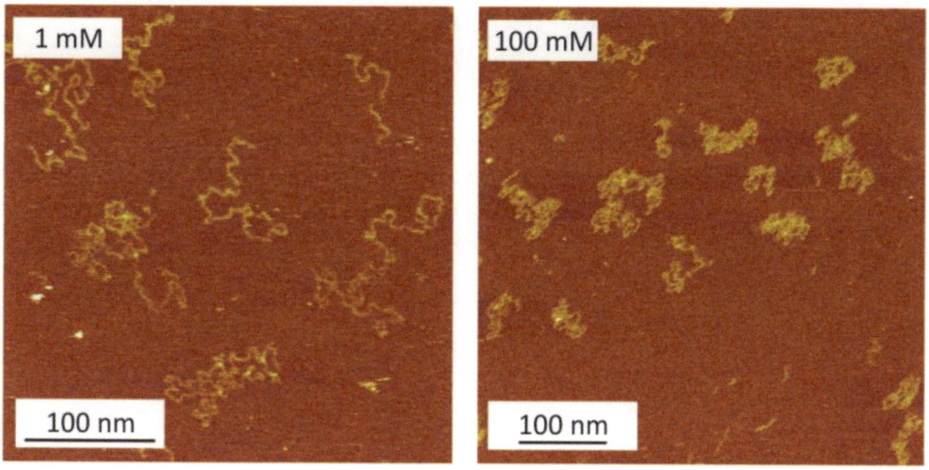

Figure 4. AFM images of poly(2-vinyl pyridine) (**P2VP**) adsorbed on mica at different ionic strength 1 and 100 mM and at pH 3.0. At this pH, **P2VP** is positively charged. At low ionic strength, the molecules form extended random coils due to intramolecular electrostatic repulsion. At high ionic strength, the electrostatic repulsion is screened, and the molecules form partially collapsed coils.

Two **P2VP** solutions were prepared by dissolving the polymer in 1 and 100 mM solutions to a concentration equal to 0.1 mg/L. At pH 3.0, **P2VP** is positively charged due to protonation of nitrogen in pyridine rings. To form a dilute **P2VP** polymer film on mica, 20 mL from 1 or 100 mM polymer solutions were adsorbed on freshly cleaved mica for 40 s. The polymer solution was then replaced with larger volume of the polymer-free electrolyte solution. The adsorbed polymer chains were imaged in amplitude-modulation intermittent contact mode. Silicon tips with nominal tip radius < 10 nm, spring constant in the range of 0.07–0.15 N/m, were used for this purpose. A scan rate of 4.88 Hz with free oscillation amplitude (FOA) of about 10 nm and an amplitude set-point of about 76% of FOA were used. The imaging was carried out at a temperature of 25°C. The image at 1 mM solution shows that the polymer chains form extended random coils on mica. This conformation is due to intramolecular electrostatic repulsion between positively charged monomers. At 100 mM, however, the polymer chains are partially collapsed. The collapse is due to screening of the intramolecular electrostatic repulsion. This observation suggests that, at the lower ionic strength, the electrostatic repulsion contributes largely to the overall conformational persistence of **P2VP** chains. Similar trends have been observed as a function of pH [2].

3. AFM force microscopy of single polymer chains

3.1. Molecular elasticity

From an analysis of the force versus extension response of single polymer chains, one may interpret their elasticity. The elasticity has two contributions: one from the loss of entropy and the other from the deformation of bond angles [23]. Bond angle deformation results in polymer length increasing beyond its contour length (the unperturbed length of polymer chain). The polymer length increases by about 10% at a force of about 2 nN [43].

The crucial step in interpretation of the elasticity of single polymer chains is the identification of single-chain responses, namely that two or more chains are not simultaneously measured. Oversight of this step would result in force responses that are stiffer than the response of an individual chain. It is equally important to ensure that the ends of the polymer chain are strongly adhered to the solid and the AFM tip; that is, the polymer does not slide over the tip or the solid. Sliding would result in softer response than the pure elastic response of the chain.

The force versus extension response is generally interpreted in terms of freely jointed chain (FJC) model [44]:

$$x = L\left[\coth\left(\frac{\ell_K F}{kT}\right) - \frac{kT}{\ell_K F} + \frac{F}{K}\right], \tag{2}$$

where L is the contour length, k is the Boltzmann constant, and T is the absolute temperature. The Kuhn length ℓ_K and the elasticity constant K represent the mechanical properties of single chains. The FJC model has been successful in the analysis of extension responses of flexible polymers, such as synthetic polymers [44].

Figure 5 shows the force versus extension responses of poly(ethylene) (**PE**), which were collected in methyl benzoate and on silica [43]. Polymer solution with concentration 100 mg/L dissolved in toluene was used for deposition. After a deposition period of about 40 s, the polymer-coated silica was rinsed multiple times with toluene to remove loosely bound polymer chains. Thereafter, repeated extension-retraction cycles of the AFM tip to and from polymer coated substrate results in the force-extension responses of single polymer chains. After modeling the individual force responses with the FJC, the extension length of each response was normalized to the fitted contour length. The figure displaying the force versus relative extension profiles shows that the responses from different chains agree reasonably well. The overlap of the profiles asserts that the responses were obtained from single chains. An average Kuhn length $\ell_K = 0.6 \pm 0.1$ nm and an elasticity constant $K = 24 \pm 3$ nN were obtained for **PE**.

Figure 6 shows the force versus relative extension responses of **P2VP** and **PS**. **P2VP** responses were collected in 1 mM, pH 3.0 solution and on mica. Sample preparation was like that explained for AFM images in **Figure 4**. Nanohandling technique was employed to ensure the placement of AFM tip on one end of the adsorbed polymers [5]. An average Kuhn length $\ell_K = 0.5 \pm 0.1$ nm and elasticity constant $K = 9.5 \pm 0.2$ nN were obtained for **P2VP**. Sample preparation in AFM measurements with **PS** was like that explained for **PE**. Measurements in solvents of different quality for **PS** show that the Kuhn length increases with solvent quality. For example, the Kuhn length increases from a value of about 0.27 nm in ethanol to a value of about 0.43 nm in toluene. This finding is akin to swelling of **PS** chains in the respective solvents. Results show an elasticity constant equal to about 21 nN, which remains the same in all solvents.

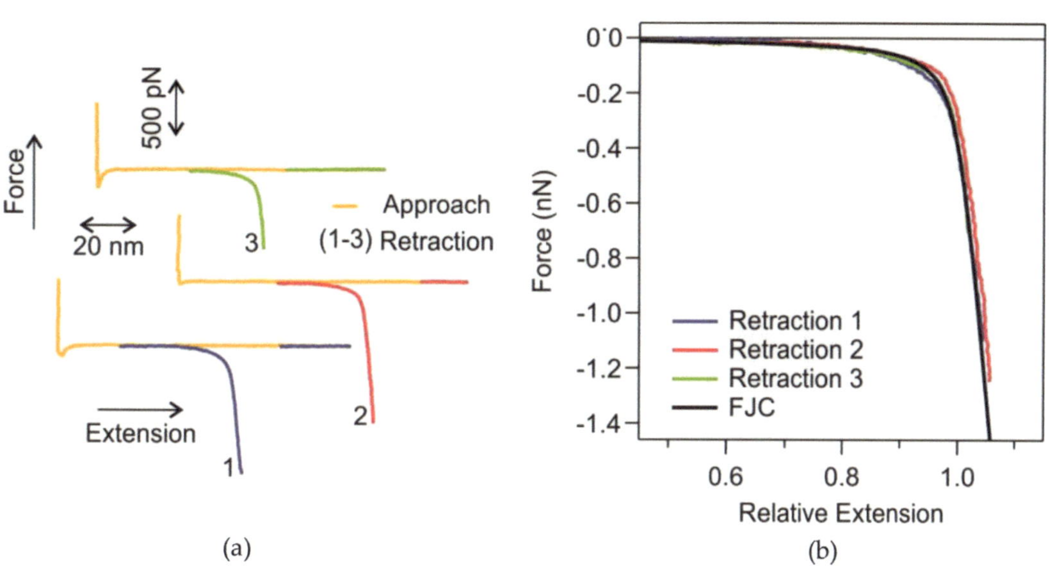

Figure 5. (a) Force versus extension responses of poly(ethylene) (**PE**) obtained from single molecule force microscopy with AFM. Measurements were performed in methyl benzoate and on silica. (b) Force versus relative extension profiles of the same retraction curves shown in (a) together with the freely jointed chain (FJC) curve. The overlap of the profiles shows that the retraction curves are responses of single chains.

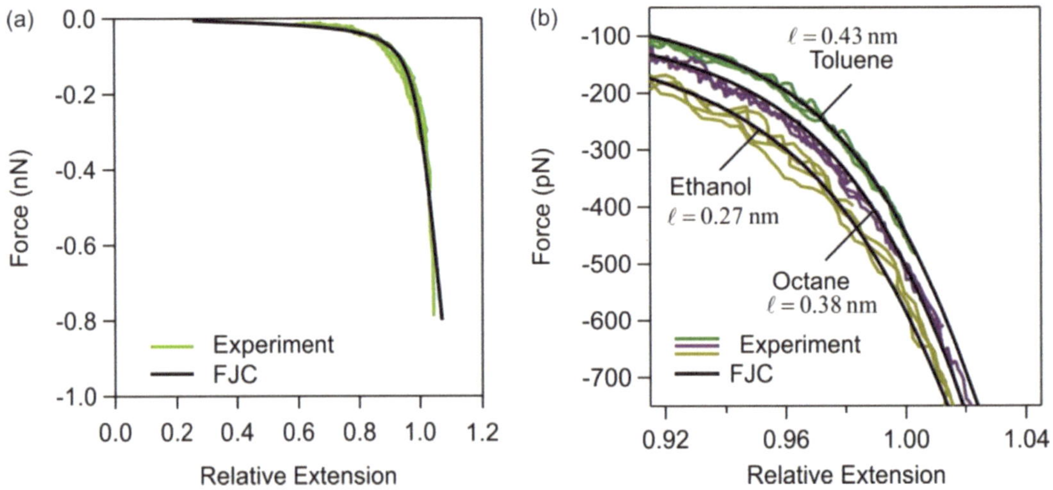

Figure 6. Force versus relative extension responses of (a) poly(2-vinyl pyridine) (**P2VP**) and (b) poly(styrene) **PS** obtained from single molecule force microscopy with AFM together with the freely jointed chain (FJC) curve. Experiments with **P2VP** were performed in pH 3.0 solution, and with **PS** in good solvents, such as toluene, to poor solvents, such as ethanol.

3.2. Adhesion force of single polymer chains

To obtain adhesion interaction forces between single polymer chains and solids, the polymer chains are generally covalently bound to the AFM tip [6, 25, 45]. The polymer chains are brought in contact with the solid. During contact, a single polymer chain may adsorb onto the solid. Upon retraction of the tip, the polymer chain desorbs resulting in a steplike (constant) force response. This force response is then fitted to a sigmoidal model giving the desorption force and length of the polymer-solid interaction.

An example of these studies is shown in **Figure 1(f)** [25]. The force versus extension response of poly(isoprene) with 88 kDa **PS** side chains in water and on hydrogen-terminated diamond shows two force response behaviors. In one case, polymer chains desorb from solid, resulting in steplike response. If two or more polymer chains desorb simultaneously, additional steps are observed in the response. Thereby, the last step is due to the final desorbed polymer chain. The second response behavior involves polymers being extended before detachment from the solid. The desorption force of polymer chains from solid may generally be tuned by the chemical environment of the polymer, polymer chemistry, and the adsorption time on the solid [6, 25].

3.3. Dynamical mechanical properties of single polymer chains

Elasticity of single polymer chains is only one property that defines their response to force. The other property is the relaxation time, or the time it takes for the polymer chain to respond to the force. Lessons from nature, e.g., wing flapping of hummingbirds, tongue projection of salamanders, or eye retraction of slugs, show that these responses are not infinitely fast but take time. This is especially important for end-tethered polymers [46].

Experiments that measure the elasticity and the relaxation time of single polymer chains generally use the thermal fluctuations of an AFM cantilever [47, 48], or externally drive the cantilever by magnetic or acoustic forces [49]. Recently, a correlation force spectroscopy (CFS) is developed that employs two AFM cantilevers in antiparallel configuration as shown in **Figure 7(a)**. The advantage of using two cantilevers in CFS, as compared with one cantilever in AFM, is that in AFM, the proximity of the cantilever to the solid increases the hydrodynamic friction due to thin film lubrication. The increase in the hydrodynamic force (or the hydrodynamic friction coefficient) increases the Brownian forces—a result of fluctuation-dissipation

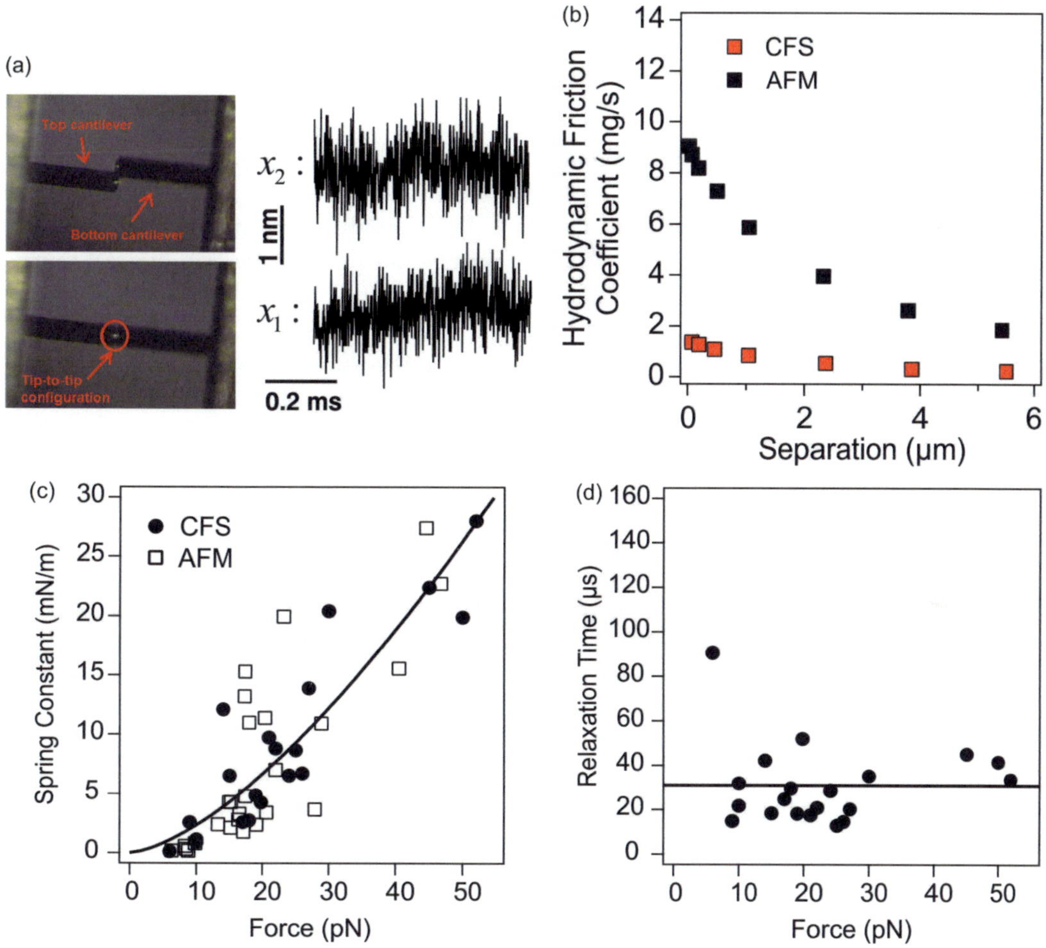

Figure 7. (a) Two AFM cantilevers in antiparallel configuration in a correlation force spectroscopy (CFS) apparatus. In the measurements, thermal fluctuations of the top and bottom cantilevers are collected simultaneously and correlated. (b) Correlation of two cantilevers' fluctuations results in a lower hydrodynamic friction in CFS than the hydrodynamic friction on a single cantilever in AFM. (c) Spring contact of single-stranded DNA measured by CFS and AFM in the force range from about 5 to 50 pN. Solid line is a fit of wormlike chain model (model may be found in Ref. [47]), resulting in persistence length equal to about 2.6 nm. (d) Relaxation time of single-stranded DNA measured by CFS in the force range from about 5 to 50 pN. Solid line is a linear fit (model in Ref. [50]), resulting in a constant value of about 31 μs for the relaxation time.

theorem [51]. Brownian forces result in thermal noise that is the major source of noise in AFM force spectroscopy measurements. Because of the thermal noise and the high hydrodynamic force, AFM force resolution is reduced, and polymer chains may only be examined accurately when extended to high forces. To reduce the high force limit, in AFM applications discussed in the above sections 3.1 and 3.2, one applies a low-pass filter to cantilever deflection signal and thereby discards the time-related or dynamical data. Placement of two AFM cantilevers in the configuration shown in **Figure 7(a)** reduces the hydrodynamic friction and the Brownian forces. **Figure 7(b)** shows a comparison between the hydrodynamic friction coefficient between AFM and CFS. In all separations (in AFM, tip-solid separation, in CFS, tip-tip separation), CFS has a lower hydrodynamic friction coefficient. Similarly, the Brownian forces or the thermal noise are lower in CFS than in AFM. Thereby, CFS has a higher force resolution. CFS also gives the dynamical mechanical properties of single molecules where no filtering is applied in the data analysis [7, 51].

In the measurements, a single polymer chain is tethered between two tips, then extended to a force and clamped. During the clamp period, thermal fluctuations of the top and bottom cantilevers are collected simultaneously. Dynamical mechanical properties of single polymer chains are obtained from an analysis of the time correlations between the two thermal fluctuations. **Figure 7(c)** and **(d)** show the stiffness and the relaxation time of end-tethered single-stranded DNA in the force range from 5 to 50 pN, respectively, [28]. One observes that the stiffness of the chain increases with the force, while the relaxation time remains almost constant equal to about 30 μs. Constant relaxation time is consistent with theory [50].

4. Mechanochemistry at the level of single polymer chains

The force versus extension response of biopolymers, such as double-stranded DNA and various proteins, supramolecules, and polymers containing force-sensitive units, namely mechanophores, generally shows a different behavior. In these polymers, specific structural changes or chemical reactions occur, which are triggered by the application of mechanical force [31–34, 52–55]. The process involves force reducing the energy barrier of transition by an amount $F\Delta x$, where Δx is a length scale associated with the transition length [56]. The reduction in the energy barrier facilitates the transition. For example, the rate of transition increases by a factor $\exp(F\Delta x)$. It has been shown that reactions that do not occur thermally may be triggered by the application of mechanical force [24, 57].

Mechanically induced isomerization of *cis* carbon-carbon double bonds to *trans* conformation in polymer **P1** is shown in **Figure 8** [37]. Experiments were realized by adsorbing polymer chains from a solution with concentration 100 mg/L dissolved in dimethyl sulfoxide (DMSO). The deposition period was 2 hr, after which the solid was rinsed multiple times with DMSO to remove loosely bound chains. Finally, DMSO was added to the solid before the measurements. **Figure 8** shows the force versus extension responses of polymer **P1**, which differ from the force behaviors of **PE**, **P2VP**, and **PS** in **Figures 5** and **6**. In the latter, the force increases with extension until the chain breaks from either the AFM tip or the solid. As shown in **Figure 8**, a

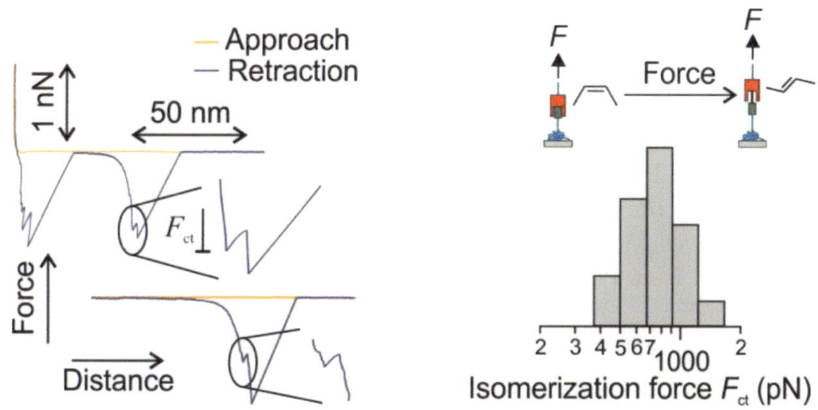

Figure 8. Force versus extension responses of polymer **P1** showing isomerization event. The onset of *cis*-to-*trans* isomerization is shown by a sudden kink in the response profile and is denoted by the isomerization force F_{ct}. The isomerization force has an average value of about 800 pN.

single chain of polymer **P1** is tethered between the AFM tip and the solid even after the isomerization. The force response of **P1** contains a sudden increase in the extension that is due to the isomerization of some *cis* monomers in the backbone of **P1** to *trans* conformation. When isomerization occurs, the force shows a sudden reduction that is due to relaxation of stress on the chain because of extension increase. The force where the isomerization occurs is denoted by F_{ct} and has an average value of about 800 pN. The isomerization force is lower than the force of breaking of covalent bonds and rings, 1–2 nN [57, 58].

5. Conclusions

AFM started as a power imaging technique and soon found its way in the diverse field of polymer studies. In this chapter, the focus was placed on those studies that are at the level of single polymer chains, that is nanoscale. AFM imaging in noncontact mode or intermittent contact mode may be used to obtain conformations and sizes of individual polymer chains. The chains ought to be adsorbed from dilute polymer solutions and on atomically flat solids. AFM force microscopy may be used to obtain the elasticity of single polymer chains. The molecular elasticity in this case is interpreted in terms of an entropic elasticity, which can be tuned by the solvent, and an elasticity term that is due to deformation of bond angles. In the case of force-sensitive polymers, AFM may be used to apply force, and thus trigger specific chemical reactions or conformational transitions in the polymer at the level of single chains and even single bonds. Technical development in AFM has resulted in techniques such as correlation force spectroscopy, which is employed to obtain the dynamical mechanical properties of single polymer chains. Finally, one should note that AFM has also been used to characterize the mechanical properties, such as adhesion, friction, and compression support, of dense polymer films and polymer brushes. This level of investigation is not single-molecule level and thereby was not included in this chapter.

Acknowledgements

Research performed by M.R. has received funding from the National Science Foundation of the United States via Award Number CBET-0959228, the National Center of Competence in Research for Bio-Inspired Materials in Switzerland, Virginia Tech, and University of Geneva. These researches were performed in the laboratory of Prof. William Ducker and in the laboratory of Prof. Michal Borkovec. M.R. acknowledges collaborations, useful discussions, and contributions from Prof. Mark Paul, Prof. John Walz, Prof. Andreas Kilbinger, Dr. Christopher Honig, Dr. Plinio Maroni, Dr. Brian Robbins, Dr. Lucie Grebikova, Svilen Kozhuharov, and Phally Kong.

Author details

Milad Radiom

Address all correspondence to: miradi@kth.se

KTH Royal Institute of Technology, Stockholm, Sweden

References

[1] Binnig G, Quate CF, Gerber C. Atomic force microscope. Physical Review Letters. 1986; **56**(9):930-933

[2] Roiter Y, Minko S. AFM single molecule experiments at the solid–liquid Interface: In situ conformation of adsorbed flexible polyelectrolyte chains. Journal of the American Chemical Society. 2005;**127**(45):15688-15689

[3] Japaridze A, Vobornik D, Lipiec E, Cerreta A, Szczerbinski J, Zenobi R, et al. Toward an effective control of DNA's submolecular conformation on a surface. Macromolecules. 2016;**49**(2):643-652

[4] Rief M, Gautel M, Oesterhelt F, Fernandez JM, Gaub HE. Reversible unfolding of individual titin immunoglobulin domains by AFM. Science. 1997;**276**(5315):1109-1112

[5] Grebikova L, Radiom M, Maroni P, Schlüter DA, Borkovec M. Recording stretching response of single polymer chains adsorbed on solid substrates. Polymer. 2016;**102**:350-362

[6] Geisler M, Netz RR, Hugel T. Pulling a single polymer molecule off a substrate reveals the binding thermodynamics of cosolutes. Angewandte Chemie International Edition. 2010; **49**(28):4730-4733

[7] Radiom M, Honig CDF, Walz JY, Paul MR, Ducker WA. A correlation force spectrometer for single molecule measurements under tensile load. Journal of Applied Physics. 2013; **113**(1):013503

[8] Netz RR. Strongly stretched semiflexible extensible polyelectrolytes and DNA. Macromolecules. 2001;**34**(21):7522-7529

[9] Dobrynin AV, Carrillo J-MY, Rubinstein M. Chains are more flexible under tension. Macromolecules. 2010;**43**(21):9181-9190

[10] Livadaru L, Netz RR, Kreuzer HJ. Interacting chain model for poly(ethylene glycol) from first principles—Stretching of a single molecule using the transfer matrix approach. Journal of Chemical Physics. 2003;**118**(3):1404-1416

[11] Hugel T, Rief M, Seitz M, Gaub HE, Netz RR. Highly stretched single polymers: Atomic-force-microscope experiments versus ab-initio theory. Physical Review Letters. 2005;**94**(4): 048301

[12] Oliveira Brett AM, Chiorcea Paquim A-M. DNA imaged on a HOPG electrode surface by AFM with controlled potential. Bioelectrochemistry. 2005;**66**(1):117-124

[13] Kiriy A, Gorodyska G, Kiriy N, Sheparovych R, Lupytsky R, Minko S, et al. AFM imaging of single polycation molecules contrasted with cyanide-bridged compounds. Macromolecules. 2005;**38**(2):501-506

[14] Lauritsen JV, Reichling M. Atomic resolution non-contact atomic force microscopy of clean metal oxide surfaces. Journal of Physics: Condensed Matter. 2010;**22**(26):263001

[15] Marchand DJ, Hsiao E, Kim SH. Non-contact AFM imaging in water using electrically driven cantilever vibration. Langmuir. 2013;**29**(22):6762-6769

[16] Grebikova L, Maroni P, Zhang B, Schlüter AD, Borkovec M. Single-molecule force measurements by nano-handling of individual dendronized polymers. ACS Nano. 2014;**8**(3): 2237-2245

[17] Adamcik J, Klinov DV, Witz G, Sekatskii SK, Dietler G. Observation of single-stranded DNA on mica and highly oriented pyrolytic graphite by atomic force microscopy. FEBS Letters. 2006;**580**(24):5671-5675

[18] Grebikova L, Kozhuharov S, Maroni P, Mikhaylov A, Dietler G, Schluter AD, et al. The persistence length of adsorbed dendronized polymers. Nanoscale. 2016;**8**(27):13498-13506

[19] Roiter Y, Trotsenko O, Tokarev V, Minko S. Single molecule experiments visualizing adsorbed polyelectrolyte molecules in the full range of mono- and divalent counterion concentrations. Journal of the American Chemical Society. 2010;**132**(39):13660-13662

[20] Kiriy A, Gorodyska G, Minko S, Jaeger W, Stepanek P, Stamm M. Cascade of coil-globule conformational transitions of single flexible polyelectrolyte molecules in poor solvent. Journal of the American Chemical Society. 2002;**124**(45):13454-13462

[21] Roiter Y, Jaeger W, Minko S. Conformation of single polyelectrolyte chains vs. salt concentration: Effects of sample history and solid substrate. Polymer. 2006;**47**(7):2493-2498

[22] Zhang B, Wepf R, Kröger M, Halperin A, Schlüter AD. Height and width of adsorbed dendronized polymers: Electron and atomic force microscopy of homologous series. Macromolecules. 2011;**44**(17):6785-6792

[23] Oesterhelt F, Rief M, Gaub HE. Single molecule force spectroscopy by AFM indicates helical structure of poly(ethylene-glycol) in water. New Journal of Physics. 1999;**1**:6.1-6.11

[24] Wang J, Kouznetsova TB, Niu Z, Rheingold AL, Craig SL. Accelerating a mechanically driven anti-Woodward–Hoffmann ring opening with a polymer lever arm effect. The Journal of Organic Chemistry. 2015;**80**(23):11895-11898

[25] Kienle S, Gallei M, Yu H, Zhang B, Krysiak S, Balzer BN, et al. Effect of molecular architecture on single polymer adhesion. Langmuir. 2014;**30**(15):4351-4357

[26] Livadaru L, Netz RR, Kreuzer HJ. Stretching response of discrete semiflexible polymers. Macromolecules. 2003;**36**(10):3732-3744

[27] Shi WQ, Giannotti MI, Zhang X, Hempenius MA, Sconherr H, Vancso GJ. Closed mechanoelectrochemical cycles of individual single-chain macromolecular motors by AFM. Angewandte Chemie International Edition. 2007;**46**(44):8400-8404

[28] Radiom M, Paul MR, Ducker WA. Dynamics of single-stranded DNA tethered to a solid. Nanotechnology. 2016;**27**(25):255701

[29] Balzer BN, Kienle S, Gallei M, von Klitzing R, Rehahn M, Hugel T. Stick-slip mechanisms at the nanoscale. Soft Materials. 2014;**12**(sup1):S106-S114

[30] Balzer BN, Gallei M, Hauf MV, Stallhofer M, Wiegleb L, Holleitner A, et al. Nanoscale friction mechanisms at solid-liquid interfaces. Angewandte Chemie International Edition. 2013;**52**(25):6541-6544

[31] Auletta T, de Jong MR, Mulder A, van Veggel F, Huskens J, Reinhoudt DN, et al. Beta-cyclodextrin host-guest complexes probed under thermodynamic equilibrium: Thermodynamics and AFM force spectroscopy. Journal of the American Chemical Society. 2004;**126**(5):1577-1584

[32] Kado S, Kimura K. Single complexation force of 18-crown-6 with ammonium ion evaluated by atomic force microscopy. Journal of the American Chemical Society. 2003;**125**(15):4560-4564

[33] Skulason H, Frisbie CD. Direct detection by atomic force microscopy of single bond forces associated with the rupture of discrete charge-transfer complexes. Journal of the American Chemical Society. 2002;**124**(50):15125-15133

[34] Embrechts A, Velders AH, Schonherr H, Vancso GJ. Self-complementary recognition of supramolecular urea-aminotriazines in solution and on surfaces. Langmuir. 2011;**27**(23):14272-14278

[35] Schuetze D, Holz K, Mueller J, Beyer MK, Luening U, Hartke B. Pinpointing mechanochemical bond rupture by embedding the mechanophore into a macrocycle. Angewandte Chemie, International Edition. 2015;**54**(8):2556-2559

[36] Klukovich HM, Kouznetsova TB, Kean ZS, Lenhardt JM, Craig SL. A backbone lever-arm effect enhances polymer mechanochemistry. Nature Chemistry. 2013;**5**(2):110-114

[37] Radiom M, Kong P, Maroni P, Schafer M, Kilbinger AFM, Borkovec M. Mechanically induced cis-to-trans isomerization of carbon-carbon double bonds using atomic force microscopy. Physical Chemistry Chemical Physics. 2016;**18**(45):31202-31210

[38] Mikhaylov A, Sekatskii S, Dietler G. DNA trace: A comprehensive software for polymer image processing. Journal of Advanced Microscopy Research. 2013;**8**(4):241-245

[39] Odijk T. Polyelectrolytes near the rod limit. Journal of Polymer Science. 1977;**15**(3):477-483

[40] Skolnick J, Fixman M. Electrostatic persistence length of a wormlike polyelectrolyte. Macromolecules. 1977;**10**(5):944-948

[41] Netz RR, Orland H. Variational theory for a single polyelectrolyte chain. European Physical Journal B. 1999;**8**(1):81-98

[42] Ullner M. Comments on the scaling behavior of flexible polyelectrolytes within the Debye–Hückel approximation. The Journal of Physical Chemistry B. 2003;**107**(32):8097-8110

[43] Radiom M, Maroni P, Wesolowski TA. Size extensivity of elastic properties of alkane fragments. Journal of Molecular Modeling. 2018;**24**:36

[44] Giannotti MI, Vancso GJ. Interrogation of single synthetic polymer chains and polysaccharides by AFM-based force spectroscopy. Chemphyschem. 2007;**8**(16):2290-2307

[45] Grebikova L, Gojzewski H, Kieviet BD, Gunnewiek MK, Vancso GJ. Pulling angle-dependent force microscopy. The Review of Scientific Instruments. 2017;**88**(3):033705

[46] Berkovich R, Hermans RI, Popa I, Stirnemann G, Garcia-Manyes S, Berne BJ, et al. Rate limit of protein elastic response is tether dependent. Proceedings of the National Academy of Sciences. 2012;**109**(36):14416-14421

[47] Khatri BS, Byrne K, Kawakami M, Brockwell DJ, Smith DA, Radford SE, et al. Internal friction of single polypeptide chains at high stretch. Faraday Discussions. 2008;**139**(0):35-51

[48] Kawakami M, Byrne K, Khatri B, McLeish TCB, Radford SE, Smith DA. Viscoelastic properties of single polysaccharide molecules determined by analysis of thermally driven oscillations of an atomic force microscope cantilever. Langmuir. 2004;**20**(21):9299-9303

[49] Kawakami M, Byrne K, Khatri BS, McLeish TCB, Radford SE, Smith DA. Viscoelastic measurements of single molecules on a millisecond time scale by magnetically driven oscillation of an atomic force microscope cantilever. Langmuir. 2005;**21**(10):4765-4772

[50] Hiraiwa T, Ohta T. Linear viscoelasticity of a single semiflexible polymer with internal friction. The Journal of Chemical Physics. 2010;**133**(4):044907

[51] Honig CDF, Radiom M, Robbins BA, Walz JY, Paul MR, Ducker WA. Correlations between the thermal vibrations of two cantilevers: Validation of deterministic analysis via the fluctuation-dissipation theorem. Applied Physics Letters. 2012;**100**(5):053121

[52] Liese S, Gensler M, Krysiak S, Schwarzl R, Achazi A, Paulus B, et al. Hydration effects turn a highly stretched polymer from an entropic into an energetic spring. ACS Nano. 2017; **11**(1):702-712

[53] Marszalek PE, Li H, Oberhauser AF, Fernandez JM. Chair-boat transitions in single polysaccharide molecules observed with force-ramp AFM. Proceedings of the National Academy of Sciences of the United States of America. 2002;**99**(7):4278-4283

[54] Clausen-Schaumann H, Rief M, Tolksdorf C, Gaub HE. Mechanical stability of single DNA molecules. Biophysical Journal. 2000;**78**(4):1997-2007

[55] Hosono N, Kushner AM, Chung J, Palmans ARA, Guan Z, Meijer EW. Forced unfolding of single-chain polymeric nanoparticles. Journal of the American Chemical Society. 2015; **137**(21):6880-6888

[56] Bell GI. Models for specific adhesion of cells to cells. Science. 1978;**200**(4342):618-627

[57] Wang J, Kouznetsova TB, Niu Z, Ong MT, Klukovich H, Rheingold AL, et al. Inducing and quantifying forbidden reactivity with single-molecule polymer mechanochemistry. Nature Chemistry. 2015;**7**(4):323-327

[58] Grandbois M, Beyer M, Rief M, Clausen-Schaumann H, Gaub HE. How strong is a covalent bond? Science. 1999;**283**(5408):1727-1730

Chapter 3

Forensic Potential of Atomic Force Microscopy with Special Focus on Age Determination of Bloodstains

Threes Smijs and Federica Galli

Additional information is available at the end of the chapter

http://dx.doi.org/10.5772/intechopen.77204

Abstract

An important aspect of any crime scene investigation is to detect, secure and analyze trace evidence. In forensic examinations where topographic characterization is important like in fingermark, textile and document forgery examinations, the atomic force microscopy (AFM) imaging technique can be of value. However, it is the force spectroscopy that could make AFM a versatile tool in crime investigations. Particularly, the ability to measure changes in mechanical properties of forensic trace material over time makes this technology in potential interesting for forensic examinations. The usefulness of force measurements to evaluate the elasticity of red blood cells (RBCs) in relation to the age of a bloodstain is an interesting example. With minimally invasive AFM technology, time-dependent alterations in the viscoelasticity of RBCs that occur during the aging of bloodstains can be featured. A discrimination between traces left by the perpetrator and other persons that have been present at the crime scene will thus be enabled. A recently obtained proof-of-concept demonstrating the usefulness of AFM for age estimation of bloodstains will be described. Additionally, the usefulness of AFM imaging and force spectroscopy for human hair, document forgery, textile fiber, fingermark and gunshot and explosive residue examinations will be discussed.

Keywords: forensic, atomic force, crime, red blood cell, textile, fingermark, explosives, forgery

1. Introduction

During a crime scene investigation (CSI), it is essential to detect, secure and interpret biological and nonbiological traces [1, 2]. The physical and chemical procedures for these forensic examinations can be performed in the laboratory but are also carried out more and more at the scene of the crime. Especially, the biological traces that can be used for DNA-profiling are important.

Crucial condition thereby is the use of non- to minimally invasive methods. Key information regarding offenses may thus be established and contribute to the reconstruction of crimes.

Today different nanotechnologies such as the application of nanoscale powders, high-resolution transmission and scanning electron microscopy (HRTEM, HRSEM) and atomic force microscopy (AFM) are available for forensic investigations [3, 4]. However, forensic trace characterization at the nanoscale may not necessarily yield relevant forensic information as explained by Inman and Rudin with the principle of divisible matter [5, 6]. Relevant forensic materials thus need to be found and interpreted preferable at the dimension at which they are created. Nanoscale, extremely detailed information may be superfluous and not related to evidence. Nevertheless, it has been demonstrated that high-resolution scanning probe AFM has interesting forensic applications [7, 8]. This chapter focuses on the forensic potential of AFM with special focus on the force measurements, known as force spectroscopy (FS). First, AFM imaging and FS are briefly explained. Second, the forensic potential of AFM imaging and FS is evaluated with special attention to age determination of bloodstains. Third, a conclusion is provided based on the effectiveness and usefulness of the information provided by this technology in relation to the system of criminal justice.

2. Principles of AFM

An AFM consists of a cantilever with a tip at the end, together named the probe, a piezo-electric XY- and Z- scanner, a laser and a photo-diode detector system (see **Figure 1**) and can be operated in different modes. In the contact mode, the tip is dragged across the surface at constant force, in the intermittent contact mode the cantilever is oscillating and the tip will be repulsed at the lowest oscillation point and get out of contact at the upper part of the oscillation, in the non-contact mode the cantilever is oscillating close to the sample but without contacting its surface and in the force modulation mode the tip is oscillating while remaining in contact with the sample surface [9].

2.1. AFM imaging

Briefly, a sample is scanned by the tip (mostly sharp and made of silicon or silicon nitride) parallel to the sample surface (with the XY-scanner) while interactions between sample and probe are experienced. These interactions concern attractive and repulsive forces between molecules of the sample and tip thereby causing deflections of the cantilever toward or away from the sample that results in a deflection of the laser beam and can be recorded by a photo-diode system (see **Figure 1**). As the deflection of the cantilever is directly proportional to the force, a feedback system is usually employed. With this feedback system, the height of the cantilever is adjusted in order to maintain a constant deflection (force) while moving parallel to the surface. In this way an image of the topography of a sample, a height image, is created, and quantitative surface roughness can be determined from height images. In another type of imaging, phase imaging, the phase shift between the driving signal and the cantilever signal as it interacts with the surface is recorded during intermittent contact AFM or noncontact

http://dx.doi.org/10.5772/intechopen.77204

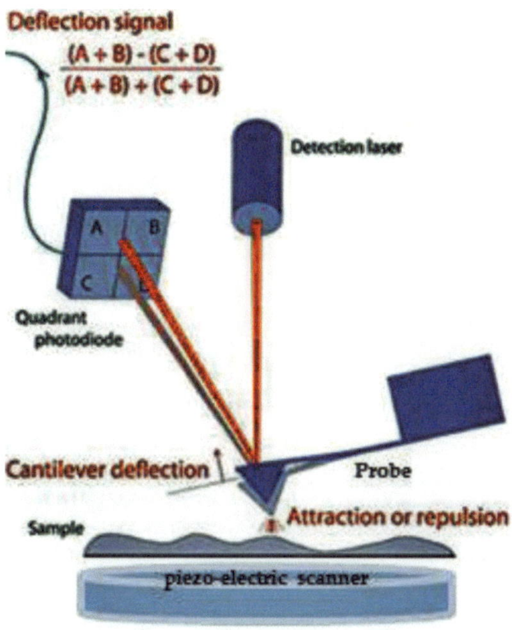

Figure 1. Illustration (kindly provided by JPK Instruments AG) of the basic components of an AFM, the probe, piezo-electric scanner, the laser, sample and photo-diode detector system. Depending on the roughness of the surface and type of measurements, the tip can be pyramid-shaped (commonly made of silicon or silicon nitride) with a curvature radius ranging from 2 nm to 2 μm or spherical (0.5–2.5 μm in diameter and mostly made of titanium or silica). Force interactions between tip and surface molecules will result in a deflection of the cantilever that is recorded as the deflection of a laser beam aligned to the back of the cantilever. Quadrant photo-diodes (with optically active areas A, B, C and D) will then convert the laser signals into an electrical output signal that is proportional to the deflection of the cantilever.

AFM. In intermittent contact, in AFM mode, the cantilever is oscillating and the tip is repulsed at the lowest oscillation point and get out of contact at the upper part of the oscillation. The phase lag is caused by the energy dissipation in the cantilever because of the experienced interaction forces between probe and sample. Dissipation differs for different materials; therefore, the phase image can provide extensive information on differences in sample composition, particularly on flat surfaces.

2.2. Force spectroscopy

In case of force measurements, the probe is moved vertically toward the sample and subsequently retracted. As the tip further approaches the surface attractive, mostly van der Waals, forces become significant. These interactions result in a "snap-to-contact" of the tip with the sample followed by a deflection away caused by repulsive molecular interactions. In this regime, the sample is indented on purpose. From the resulting force-distance curve (FD, see for a schematic representation **Figure 2**), cantilever properties and contact area nanomechanical properties such as a material's modulus of elasticity, the Young's modulus (YM), can be quantitatively obtained. The YM is a mechanical property that indicates the force per unit area that is needed to compress or stretch an elastic material. Stiffer materials have larger YM. This contact stiffness can

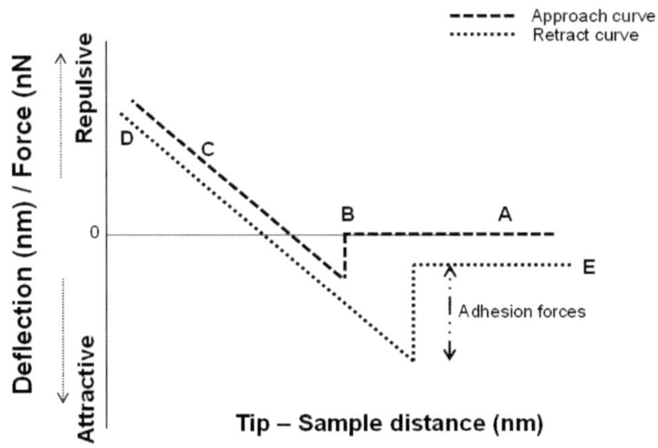

Figure 2. Schematic representation of idealized FD curves showing the approach and retract curve. A: the tip is approaching the sample surface; B: van der Waals attractive forces are experienced resulting in a snap-to-contact of the tip with the sample and indentation; C: deflection of the tip away from the sample; D: retraction of the tip; E: withdrawal of the tip from the sample. The larger the deflection of the cantilever, the stiffer the sample (see the slope of part C).

be retrieved from the FD curves that can be measured at slow rate [force-volume (FV)] [10], at high rate [11], via pulsed-force mode [12], peak force [13] or from the change in resonance frequency of the cantilever (contact-resonance AFM) [14] and by amplitude modulation [15]. Other mechanical properties that can be derived from FD curves are visco-elasticity, adhesion forces and energy (the area of the attractive part of the retract FD-curve) [16]. Additionally, FD-curve modifications induced by electrostatic charges can be investigated [17]. Electrical signals based on the conductivity through sample and tip may be monitored as well while the tip is moving over the sample [9]. It may be noticed that for mechanical studies with lateral resolution a sharp silicon or silicon nitride tip can be used but a spherical, colloidal probe may render more specific data on material-material mechanical interactions like forensically relevant adhesion interactions between gunshot and explosive residues and textile fibers [18].

3. Forensic potential of AFM imaging

As demonstrated in several studies, little research has been carried out into the forensic possibilities of AFM imaging [7, 18]. Subjects that have been investigated concern human hair analysis [19–24], document forgery [25–28], textile fibers [29, 30], fingermarks [31], gunshot and explosive residues [32, 33]. It is important to discriminate between those studies that have actually used this technique from a forensic point of view from those that employed AFM only to examine materials that may be present as traces at a crime scene. In case of forensic traces relevant micro- to nanosized regions are not visible at macro-level and thus difficult to detect. Even though AFM is a minimally invasive technique, an important condition for the preservation of evidence, the small scan area of usefully not more than 100 × 100 μm strongly limits a forensic application. In this respect, HRTEM/SEM mediated investigations could be more useful if it were not that the required sample treatments for electron microscopy imaging have been proven responsible for surface modifications of the samples [34]. Also, electron microscopes operate in the majority of cases in high vacuum conditions. Atomic force microscopy does not require any specific form of sample preparation and operates in ambient conditions. In this section, important (potential) forensic applications of AFM imaging will be introduced and analyzed based on key articles from the field. In comparison with other currently available techniques, the value of AFM imaging to a CSI will be explored.

3.1. Hair analysis

Forensic hair analyses may be used for genotyping but also to provide information on past drug exposure. Current forensic hair analyses are mostly performed with ultra-high performance liquid chromatography–tandem mass spectrometry (UHPLC–MS/MS). Most of the general hair examinations with AFM imaging focused on the influence of chemicals, as for instance present in hair care products, on the surface structure of hair. Durkan *et al.*, for instance, showed a reduction in hair surface roughness after washing with a number of shampoos to typically below 10 nm (AFM Cambridgenano CN6000 SPM). This reduction was proven to be directly related to the type of product used [19]. In some cases, isolated deposits were left behind but how long they remained visible was, unfortunately, not investigated. Actual toxicological hair investigations using AFM imaging have not been reported so far. However, unless combined to Raman spectroscopy AFM may not be able to give a chemical identity of drugs or other hair deposits and therefore just play a minor role in toxicological hair investigations. Another difficulty of AFM imaging in a forensic hair examination is lack of a reference hair sample with representative physical properties. Variations in characteristics depend on the origin of the hair, the region within the origin (thus the age of the hair), differences in the hair producing follicles, environmental conditions and personal care habits. This subject has also been acknowledged by Gurden *et al.* in 2004 (AFM NanoScope IIIa; spring constant 0.06 N/m; loading force 3.6 nN) who tried to solve this problem with a classification of hair properties based on several cuticular descriptors calculated from the height images of various hair parts [21]. These cuticular descriptors provide a range of information on hair surface properties, and thus the possibility to correlate hair structure characteristics to environmental conditions the hairs have been exposed to. The forensic relevance of this has never been established but probably requires a more extensive database including not only imaging data but mechanical hair properties as well. Jeong *et al.* have given an interesting contribution to this subject by studying the effects of aging on normal Korean hair diameter and surface features with AFM (AFM NANOStation II; non-contact mode; frequency 146–236 kHz; spring constant 1–98 N/m) [23]. Any information that may contribute to estimating the age of a forensic trace is extremely valuable. The value of this lies in the fact that a trace deposition time can link a suspect to the time a crime has been committed. Interestingly, Jeong *et al.* discovered an increase in hair diameter in the first 20–30 years followed by a decrease with further age increase. For the cuticular descriptors, surface roughness increased also significantly with age. However, the presented results showed a large variation and thus low precision. It may be noticed that gender had no influence on the hair diameter of the Korean participants. The hair surface area studies of Tomes *et al.* with both SEM and AFM (AFM

Veeco; resonance frequency 300 kHz; spring constant 42 N/m) resulted in only little difference in quality of surface profiles obtained with these techniques [22]. For forensic hair imaging, the minimally invasive AFM technique may thus be preferred over SEM even though large hair surfaces may limit its effectiveness. The forensic relevance of the preservation of evidence weighs in this case heavier.

3.2. Document forgery

Document forgery involves the illegal altering, erasing or extension of its contents. The minimally invasive character of AFM imaging when compared to SEM fits well to this subject. Competitive less invasive techniques in this case are FTIR, Raman spectroscopy, near infrared hyperspectral imaging and digital technologies [35–37]. With respect to this subject, it is interesting that the study of Kasas *et al.* (Nanoscope IIIa) on line crossings produced with dot matrix printers and different ball-point pens on plain paper reported qualitatively the same or better AFM results when compared to SEM. Similar superior AFM quality over SEM in crossing line investigations has been reported by Chen *et al.* [27]. They demonstrated different roughness and maximum height values for commonly used paper (duplicator, copper printing, glassine and Kraft paper) samples whereby crossing lines were applied with three different types of oil-based pens (AFM Bruker; resonance frequency: 146–236 KHz; spring constant: 21–98 N/m). However, the number of scans in the 5×5 μm areas and the number of these spots were not given and significances in paper surface roughness could therefore not be given. Additionally, the authors provided amplitude images. These types of images normally show how the tip is deflected when encountering the sample's topography while the feedback system is trying to keep the amplitude constant. Because the deflection and amplitude images are actually the error signals, good amplitude (deflection) images will only be obtained in case of minimized deflection signals. Although this research proved the usefulness of AFM imaging to detect crossing lines in general, the overall paper surface roughness could hamper the detection of erased, partially erased streaks or slightly printed ink patterns. It is also important to realize that changes in height profiles of ink streaks on documents may result from absorption of the ink by the paper substrate. The impact of this phenomenon may vary for different types of paper and hinder a correct interpretation of the height images. Moreover, when only AFM imaging is applied in these types of investigations a clear evidence of counterfeit can never be provided. A chemical identification of the ink with for instance Raman spectroscopy is thereby indispensable to deliver the final crucial decisive information. A combination of AFM and Raman spectroscopy has therefore more forensic potential as demonstrated by Bradao *et al.* [28]. This investigation focused on the authentication of banknotes (US dollar, Euro and Brazilian real) on the basis of asymmetry and kurtosis for the evaluation of paper roughness. Based on these parameters AFM (spring constant 42 N/m, resonance frequency 285 kHz) could, in most of the cases, discriminate the paper that was used for the counterfeiting compared to that of the authentic banknote (based on scans from different banknote locations). As stated by Ellen, AFM imaging may provide useful information on crossing lines and thus on document forgery and the order of text application [26]. However, further research is warranted to confirm the same degree of usability for rougher paper or documents that have been exposed to extreme environmental conditions. Moreover, current optical document examinations also focus on the chemical identity of crossing. The chemical heterogeneity that could be present in fraudulent documents could be well characterized with AFM phase imaging particularly in combination with infrared or Raman spectroscopy for chemical ink analysis.

3.3. Textile fibers

Current forensic textile investigations comprise microscopic, chemical and/or mechanical analyses. Although the value of forensic textile investigations concerns more, the mechanical properties of textile fibers under various environmental and weathering conditions [29]. AFM height images may deliver height, valley and mean square roughness values for different types of textiles. This subject has been investigated by Canetta *et al.* (JPK Bio AFM, Au-coated Si_3N_4 cantilever, spring constant: 0.03 N/m) in a study on environmentally stressed and weathered textile fibers [30]. An important result of this research is that the surface roughness differed significantly for all investigated textile fibers (natural cotton, wool and man-made viscose) and that it showed a time and environment dependent increase. Atomic force microscopy imaging may thus be used to distinguish the effect of different environmental effects on fibers and is thereby complementary to other microscopy techniques like scanning SEM and environmental scanning electron microscopy (ESEM). However, a forensic textile investigation requires also information on the chemical identity of textile fibers. Additional noninvasive techniques such as surface-enhanced Raman scattering, Raman microspectroscopy, FTIR or photodiodearray spectrophotometry are therefore still needed. As these techniques can provide both nature and color of the textile fibers [38], AFM imaging will not play a crucial role in forensic textile investigations. Unless combined to Raman spectroscopy, AFM imaging can only add complementary information on textile characteristics and degradation patterns.

3.4. Fingermarks

A fingermark is an impression of friction ridges of a human finger that consists of exogenous and endogenous compounds. In a CSI visualization of latent fingermarks is important because the patterns of ridges of a human fingermark are very characteristic and therefore a powerful biometric feature for a person's identification. A chemical identification of the fingermark components and their metabolites render additional donor information such as personal habits and health condition. The various physical, chemical and instrumental techniques [39–43] for fingermark visualization and analysis focus on improving the contrast between the ridges and the surface underneath, surface characteristics and the presence and identification of particular contaminants [44, 45]. The subject of age determination of a fingermark is also with respect to forensic fingermark research an important but not yet solved issue [46–48]. Atomic force microscopy imaging could highlight specific details of fingermark ridges and substrate surface provided that the roughness of the surface on which the fingermark was deposited does not increase the height of the fingermark ridges. This problem was indeed experienced in the study of Goddard *et al.* (AFM Veeco; resonance frequency 250–350 kHz; spring constant 20–80 N/m) [31]. In their study that focused on localized brass surface erosion processes in relation to fingermark ridge evaluation, the authors noticed how the high roughness of the brass surface in a 70 × 70 μm scan area hampered the analysis of the ridges of a fingermark. In this particular research, these problems were solved by successfully polishing the brass surfaces. However, as stated by the authors themselves, this solution is far from realistic. Apart from the obstacle the surface roughness is giving, the inability of AFM to provide chemical information on the components of a fingermark further limits its forensic usefulness. Spectroscopy techniques that operate in a nondestructive manner seem to have more value in case of forensic fingermark examinations. Vibrational spectroscopy techniques like FTIR and conventional Raman are examples of methods that offer in a nondestructive manner a specificity for molecular identification that is comparable to mass spectrometry [41, 43]. The latter has nevertheless excellent selectivity and sensitivity in identifying unknown fingermark components but is highly destructive. Even AFM phase imaging cannot offer similar forensic effectiveness. This type of imaging can, unhindered by surface roughness, admittedly provide information on different materials and provide physicochemical mapping of exogenous substances present on fingermarks (for instance, gunshot or explosive residues or compounds that can be associated with sexual assaults) but fails in chemically identifying them. There is, however, still one application of AFM imaging that remains to be explored but could be potentially interesting in a forensic examination, the investigation and deconvolution of overlapping fingermarks and/or bloodstains.

3.5. Gunshot and explosive residues

Inorganic gunshot and explosive residues can provide important information in the forensic reconstruction of shooting incidents and are usually analyzed with neutron activation analysis (NAA) [49], atomic absorption spectrometry (AAS)-based methods [50, 51], inductively coupled plasma (ICP) [52], and SEM combined to energy dispersion analysis (SEM–EDX) [53]. Organic gunshot and explosive residues can be analyzed with gas chromatography (GC), GS-MS, or HPLC [54]. Additionally, time-of-flight secondary ion mass spectrometry (TOF–SIMS), ablation-ICP/MS and Raman micro-spectroscopy have been reported for both inorganic and organic gunshot and explosive residue characterization [55, 56]. To clarify the chemical identity of these residues, it is an extremely important element in CSIs that involve firearms. Consequently, AFM imaging will only have any value when combined to one of the indicated techniques. Such a combination of technologies has been reported to be successful to evaluate shooting distances based on the shape and size of GSR (Quesant Q-Scope 250 Nomad) [57]. D'Uffizi *et al.* explored micromechanical and micromorphological features of gunshot residue particles deposited on bullets and hands of a shooter and collected with double-sided tape. In combination with SEM-energy-dispersive spectroscopy and selected-area X-ray photoelectron spectroscopy, the role of AFM height and phase imaging in this investigation was only modest [32]. The only forensic gunshot and explosive investigation in which AFM imaging could provide sufficiently powerful information regards physicochemical characterization of gunshot and explosive residues that have been detected on hairs and in-between the ridges of fingermarks as demonstrated in several publications [21, 33]. In this respect, the study of Oxley *et al.* (Digital Instrument Dimension 31,000) proved that various chemical hair treatments (water, acetonitrile, KOH and $KMnO_4$) all resulted in a decrease in surface roughness. However, no information was given on the recovery time, thus the duration of the observed decrease in roughness [33]. Moreover, finding the right location will still be difficult and time-consuming.

4. Forensic potential of force spectroscopy

In contrast to the limited value of AFM imaging in a CSI, FS can in fact play a crucial role in these investigations. The power of this AFM application concerns the possibility to measure forces at the nano-level and consequently calculate mechanical characteristics of materials and changes of these properties over time. Changes in mechanical characteristics of forensic trace material as function of time render the possibility to estimate the age of the trace. Yet there is little forensically mediated research on this subject. Investigations in this area have mainly focused on age determination of bloodstains [58–61] and to a lesser extent on hair [19, 20, 23, 62, 63], gunshot and explosive residues, [64] and pressure sensitive adhesives [65–67]. Based on key articles from the field, the application of FS in forensic hair, gunshot and explosive residue and pressure sensitive adhesives examinations will be discussed followed by more extensive attention to age determination of bloodstains.

4.1. Hair

As demonstrated by Durkan *et al.* force measurements (AFM Cambridgenano CN6000 SPM; spring constant 0.2 N/m) proved 20% increase in adhesive force and four times higher adhesion energy on hair deposits as compared to the bare hair surface with 6 nN adhesion force. The hair deposits in this study resulted from washing of the hairs with various shampoos. No data were available on statistical significance of these findings. More forensically interesting is the study of Jeong *et al.* that demonstrated the usefulness of AFM force measurements by showing age-dependent hair stiffness. The stiffness increased up to 30 years and then decreased again. The average adhesion force of the hairs, however, showed no age dependence. As also noticed by the authors, their data may not be representative for all hairs and hair parts. Force-distance and friction measurements between individual hair strands have been documented by Max *et al.* for direct quantification of hair-hair (other types of fiber) interactions in the range of 10–100 mm diameter (AFM Atomic Force F&E, MFP-3D; spring constant 1.2–8.5 N/m) [62]. DelRio and Cook have also provided interesting data of hair samples (untreated virgin hairs and conditioned and bleached hairs) [18]. They reported an indentation modulus of 2.4 ± 1.1 GPa and 1.8 ± 0.9 GPa for respectively the virgin and the bleached hairs. For the conditioned hairs, the modulus of indentation varied between 0.05 and 0.5 GPa depending on the position along the hair. All the measurements were performed on a 5 by 5 μm area but the number of indentations was not mentioned. Although all these investigations are interesting from a physical point of view, the forensic usefulness of mechanical hair parameters will be limited because of the numerous environmental and personal conditions affecting these parameters and thus the accuracy of the overall measurement. However, modeled in a Bayesian network, mechanical properties may be related to environmental conditions and used to calculate complex likelihood ratios and thus increase their usefulness in a CSI.

4.2. Gunshot and explosive residues

Apart from the topography of organic and inorganic particles present in gunshot residues and explosives, mechanical properties of these particles have a forensic value as well. With respect to this subject, Xu *et al.* have described an interesting nanoscale characterization of mock explosive materials, a study that fully explored the technical options of dynamic AFM such as phase imaging, force volume imaging, and Kelvin probe force microscopy (KPFM) with resonance enhancement (AFM Asylum Research MFP-3D AFM; spring constant 10.97 N/m; resonance frequency 147.53 kHz) [64]. Physical properties of components of explosive residues such as the density of the simulant, simulant to the explosive HMX and polymeric binder were mapped over 10×10 μm areas to understand the formation of hotspots and their local structure in relation to the processing method. This enabled mapping of local mechanical dissipation, elastic modulus, adhesion and the "effective" local dielectric constant of a mock explosive 900-21 sample (mechanical substitute for the plastic-bonded explosive PBX 9501). The authors used the phase lag between the excitation force and tip response for a nanoscale quantitative analysis of their sample and emphasized the importance of measuring this phase lag. To make the phase images more useful for quantitative mapping, conversion into energy dissipation maps was included. If the amplitude of the cantilever is kept constant, the phase shift is related to the energy dissipated in the tip-sample contact and phase data can thus provide energy dissipation maps [68]. The energy dissipation during one oscillation cycle by the tip on the sample was calculated according to the well-established method described by Martinez and Garcia [69]. To create adhesion and YM maps, FV mapping was employed based on Hertz contact model calculations. When compared to energy dissipation, adhesion or YM maps, the dielectric property map revealed more localized spatial features: fine and large crystals appeared to have the same dielectric constant while the binder region was characterized by a much higher dielectric constant. This study clearly demonstrated AFM multiparameter functionality resulting in a variety of physicochemical properties of compound mixtures. Particularly, the interfacial regions between crystalline zones of complex composite materials such as explosive residues were indicated as important areas where impurities, unreacted molecules, additives and binders form a heterogeneous structure. The study of DelRio and Cook also investigated adhesion forces of explosive particles on different fabric types using a colloidal probe [18]. The results showed for two fabrics similar modulus of indentation (29.0 ± 8.0 MPa for cotton and 30.7 ± 7.0 MPa for rayon) that differed from the values given in the literature (3 and 11 GPa for respectively cotton and rayon fibers) [70]. The authors ascribed the differences to variations in surface roughness and work of adhesion. This example illustrates the difficulty in quantifying mechanical properties based on AFM and determining the "true" value of a mechanical property. Some of the choices that can be made for in a given AFM measurement and the probes may influence the outcome of mechanical measurements and thus need to be correctly interpreted to insure the forensically desired accuracy and precision.

4.3. Pressure sensitive adhesives

Pressure sensitive adhesives (PSA) usually consist of a polymeric base with appropriate plasticizers and tackifiers and are, unfortunately, also employed in criminal activities such as assault, rape and hijacking to immobilize and blindfold victims and in the preparation of homemade explosives. Typically, PSA investigations involve physical fiber characterization and chemical analysis of the adhesive additional to the search for fingermarks and DNA. A physical fit of free tape ends may constitute important evidence in the reconstruction of a crime and link a suspect to a scene of crime [71]. Current methods for PSA examinations of physical characteristics include polarized light microscopy, GS-MS [72], FTIR [72, 73], SEM [74], X-ray fluorescence spectrometry (XRF) [75], and ICP [76]. As PSA manufacturing and effectiveness depends on the cohesive and viscoelastic properties of the applied polymers, AFM may certainly play a role in forensic PSA examinations. In 2000, Paiva *et al.* applied AFM (AFM Autoprobe M5) in a PSA investigation for surface adhesion measurements on 7 months old polyethylene-propylene blends with different concentrations of the tackifier n-butyl ester of abietic acid [65]. A two-phase morphology was found for the blends at compositions exceeding 15 wt% tackifier. This resulted in two different types of behavior as demonstrated via nanoindentation measurements: a viscoelastic response in the tackifier-rich domains versus a more highly dissipative response in the matrix. In 2001, the same group used AFM to investigate PSA aging [67]. Apart from the polyethylene propylene/n-butyl abietate blends combinations of polyisoprene/n-butyl abietate and polyisoprene/pentalyn H were investigated after 2 weeks and 18 months with various tackifier concentrations. The microindentations in their study used a glass hemisphere probe (3 mm in diameter) and a maximum load of 25 mN while the nanoindentations were performed with a conical tip with a curvature radius of 10 nm. Adhesion measurements provided information on blend changes over time. While these changes appeared to be significant for the miscible polyisoprene/n-butyl abietate and polyisoprene/pentalyn H systems adhesion properties were not affected for the immiscible polyethylene propylene/n-butyl abietate samples. Moreover, adhesive stiffness revealed a more pronounced increase (elasticity reduction) as function of the tackifier concentrations in the miscible systems than in the immiscible system. This stiffening degrades the adhesive effectiveness of the films that could prove an interesting feature in a forensic investigation. Canetta *et al.* also presented different nanostructural and nanomechanical properties of a variety of adhesive tapes: three visually indistinguishable different, transparent OPP packaging tapes, the visually distinguishable brown packaging, and green electrical insulation tape (AFM NTEGRA; spring constant 45 N/m; resonance frequency 330 kHz) [66]. All tapes showed the existence of two phases: a hard low energy dissipative phase and a soft phase which exhibits more energy dissipation. The authors correctly noticed that in case of adhesive tapes with dispersed regions of higher and lower viscosity, the interaction of the AFM tip with the surface may vary. Consequently, less energy will be dissipated on surface regions with lower viscosity while in the more viscous areas enhanced energy dissipation occurs. For both the visually distinguishable and indistinguishable tapes, AFM measurements demonstrated statistical differences in the maximum adhesive force of the particles to the tip, the maximum distance of deformation of these particles and the adhesion energy. Atomic force microscopy can certainly give relevant nanomechanical information in a forensic PSA investigation that may not easily be clarified with other available techniques provided that the effect of environmental conditions on PSA is documented as well. Paiva *et al.* have also made a start with this and proved a decrease of elastic modulus with increasing relative humidity, estimated with Hertzian contact mechanics (Autoprobe CP scanning probe microscope operating in lateral force microscopy mode using the signal access module) [65].

4.4. Age determination of bloodstains

The most interesting forensic application of FS concerns undoubtedly the estimation of the age of a bloodstain at a crime scene. The need for estimating the deposition time of a trace of blood, linking a suspect to the time a crime has been committed, has attracted much attention worldwide over the years. Although many studies have focused on blood as the key trace at a crime scene, it is currently still not possible to provide an accurate determination of the age of a bloodstain. Except for hyperspectral imaging (HSI), technologies explored were all characterized as invasive and thus less attractive for forensic applications. Even though HSI is a promising technology deviations of the true age of bloodstains compared to the age determined with HSI already increase within a few days (for an actual age of 2 days, the absolute error is 2.7 days) [77]. An innovative and minimally invasive method for age determination of blood traces at a crime scene is thus highly needed. Force spectroscopy could provide estimations for bloodstain ages based on temporal changes in the elasticity of red blood cells (RBCs) [7, 8, 58–61]. One of the first FS researches related to this subjects was performed in 2007 by Strasser *et al.* (AFM Topometri Accurex; spring constant 80 N/m; resonance frequency 405 kHz) [59]. However, indentation areas were not correlated to specific RBCs, and thereby the condition of elasticity also neglected. On the contrary, the more extensive study of *Wu et al.* (AFM Veeco Autoprobe CP Research; frequency 72–96 kHz; spring constant 3 N/m (tapping mode); frequency 255–315 kHz; spring constant 0.9 N/m (contact mode)) showed a time-dependent, surface-dependent (glass and mica) and temperature-dependent (controlled 25°C, 76% humidity vs. uncontrolled outdoor 21–34°C, 38–87% humidity) increase of surface adhesive forces of RBCs between 5 and 30 days [58]. This result is indicative for an increase in stiffness of these cells over time. Chen and Cai (AFM AutoProbe CP; spring constant: 2.8 N/m) also observed time-dependent cellular changes in blood cells on a mica carrier in air [61]. Apart from the effects of temperature and humidity on the elasticity of RBCs, the influence of drugs has been investigated as well. The extensive study of Girasole *et al.* (contact mode spring constant 0.032 or 0.064; tapping mode spring constant 5 N/m) demonstrated that in vitro treatment of RBCs with the drug nifedipine, used in cases of cardiovascular disorders, caused dramatic morphological cellular changes that depended predominantly on the drug concentration and to a lesser extent on the exposure time [60]. The YM values calculated for the nifedipine-treated cells compared to control, native dehydrated RBCs, phenylhydrazine and formalin treated cells were respectively 96±14 kPa, 98±17kPa, 150±23kPa and 191±28kPa. Additional to the YM and shape of RBCs the authors emphasized the importance of other smaller, though significant age markers that can be found on the surface of RBCs such as, for example, spicules and crenatures. The YM measurements in this study were carried out carefully using the Hertz model adjusted for a four-sided pyramidal shape of the indenter as given by Sneddon [78] and Bilodeau [79] (see Eq. 1), a maximum load of close to 2.5 nN and an observed indentation depth of 50–60 nm.

$$F = \frac{C_0 2\delta^2 E}{\pi(1-\nu^2)\tan\alpha} \tag{1}$$

In this equation, F is the applied force (N), E is the YM (Pa), ν is the Poisson's ratio (typically 0.5 for elastic bodies), α is the apical tip angle and the coefficient, C_0 (1.46) is the specific contribution of Bilodeau in the case of pyramidal tips. After the study of Girasole in 2012, it still took approximately 5 years before Smijs and Galli paid serious attention to the age determination of bloodstains using FS (AFM JPK instruments AG; frequency 300 kHz; spring constants 25.2–67.5 N/m) [8]. They applied FS combined to the Hertz model (see Eq. 2) to investigate the elasticity of randomly selected RBCs from the peripheral zone of 4- to 8-day-old bloodstains under controlled laboratory conditions.

$$F = \frac{4}{3}\frac{E}{1-\nu^2}\sqrt{r}\cdot\delta^{\frac{3}{2}} \tag{2}$$

In this equation, F represents the applied force (N), r the tip radius (m), E the Young's Modulus (Pa), ν the Poisson ratio and δ the deformation at maximum load (m). Special attention was paid to the condition of the silicon probes when continuously used to indent RBCs. An important conclusion of this study was that, based on 256 × 256 indentations/RBC, the elasticity of six RBCs from a 5-day-old bloodstain appeared homogenous over the cell (see **Figure 3A**) with a mean Young's modulus of 1.6 ± 0.4 GPa (see **Figure 3B**). Differences in RBC YM were significant but because of the large number of YM, data significance is here mainly caused by the size of the sample rather than by the chosen level of significance and thus leading to the detection of extremely small differences. Moreover, the authors showed that the eta-squared (η^2) effect size appeared to be 0.065. This means that the spreading between the data was for only 6.5% caused by the RBC itself. At the same time, this finding illustrates the complexity of age determination of a bloodstain in a forensic setting where a variety of environmental as well intrinsic blood and bloodstain factors affect the stiffening of RBCs.

Then authors noticed extremely blunting of their silicon tip resulting from many RBC indentations (from the original radius of 8–200 nm after a total of two sapphire calibrations and six RBC (256 × 256/cell) indentations. The contact area thus changes resulting in differences in physics at that nano-level. The authors solved this problem by checking the tip's radius before

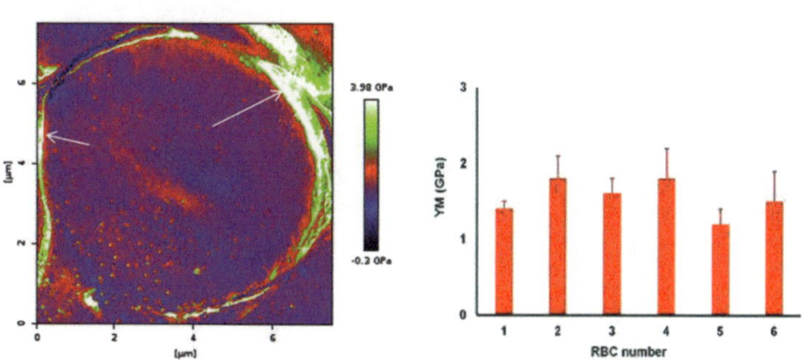

(a) (b)

Figure 3. A representative example of an elasticity map (A) of a RBC and the mean YM (± SD, B) calculated for six RBCs. All RBCs were randomly chosen from the peripheral zone of a 5-day old bloodstain (6.5–7 mm in diameter, obtained from 3 μL capillary blood). The arrows in A indicate areas with artifacts probably based on a sharp fall of the tip at the edge of the RBC. The blood drop was passively deposited on a glass surface in a Petri-dish, dried (23.9 ± 0.5°C and 35 ± 7% relative humidity, n = 7) and measured after 5 days (27°C, 36% relative humidity). AFM (JPK Instruments AG) specifications: sapphire calibrated silicon tip (Olympus), spring constant 38.75 N/m; frequency 300 kHz; maximum load: 548 to 874 nN; indentation depths: 12 to 16 nm, [8]. Copyright © 2017 Smijs T.

and after RBC indentations. Up to a radius increase to approximately 100 nm, corrected radii were applied to the Hertz model. However, the authors rightfully recognized the shortcomings of this technique when it comes to an application on a real crime investigation with an unknown bloodstain with RBCs of unknown stiffness. A problem related to this, and also acknowledged by the authors, is the fast increasing stiffness of the RBCs over time and thus the need for an ever-stronger cantilever in order to assure indentation. These things were particularly noticed when RBC stiffness's of a bloodstain aging between 4 and 8 days was investigated (see **Figure 4A** and **B**).

AFM specifications for A: sapphire calibrated silicon tips (Micromash) with spring constants ranging from 45.5 N/m, 55.8 N/m and 55.1 for respectively day 4, day 5, 6 and day 7, 8; frequency: 300 kHz; maximum load: 3018 –3414 nN; indentation depths: 20–45 nm. Statistical data analysis (A) was performed with One-Way Repeated Measures Anova (IBM SPSS statistics 20) with a critical level of significance of $p = 0.05$ and based on YM values calculated with the Hertz model (Poisson ratio: 0.5; corrected radii: 20 nm for day 4 data, 70 nm for day 5 data, 133 nm for day 6 data, 75 nm for day 7 data, 130 nm for day 8 data) for 43,046 to 44,272 FD curves per cell. Given are the mean YM values/cell (± SD). Size effect η^2 for the factor day is 0.810. Calculated size effects for day 5 vs. 6, 6 vs. 7, and 7 vs. 8 were respectively 0.195, 0.727 and 0.597.

Presented mean YM in B (Hertz model, Poisson ratio: 0.5) were based on force measurements of three RBCs (equality variance proven, Mauchly's test) from the peripheral zone of the stain. AFM specifications: sapphire calibrated silicon tips (Olympus and Micromash) with spring constants of 25.2–67.5 N/m; frequency: 300 kHz; maximum load: 548–3905 nN; indentation depths: 8–63 nm.

Similar to other researchers, Smijs and Galli also found irregular values for the YM of RBCs between 2 and 4 days old. Wu *et al.* ascribed these fluctuations to a collapse of the cell, a feature that was also noticed by Girasole *et al.* Moreover, both Smijs *et al.* and Girasole noticed stiff spicules and crenatures on the surface of the RBCs [8, 60]. Smijs and Galli concluded from their study that additional experiments using similar but also realistic forensic conditions with

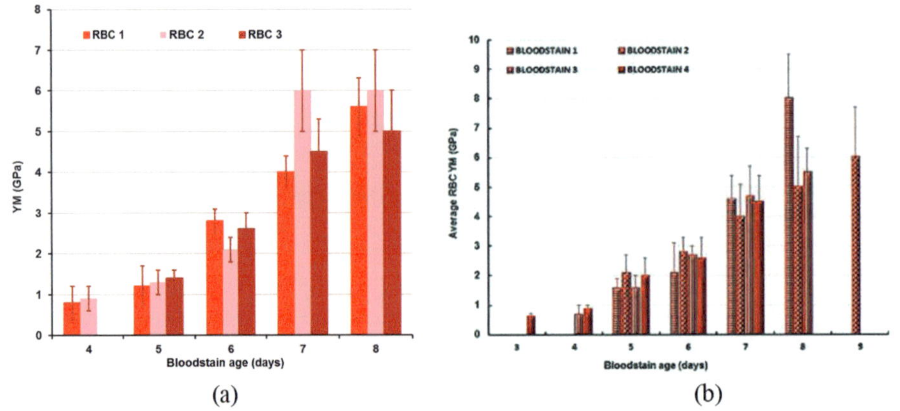

Figure 4. Representative changes in YM (± SD) of RBCs selected from the peripheral zone of a bloodstain between 4 and 8 days old (A) and changes in elasticity (YM ± SD) of cells from different 3- to 8-day-old bloodstains (B) [8]. Copyright © 2017 Smijs T.

optimized AFM probes, such as those with a more robust tip, so that the tip radius remains constant during the measurements, are needed. As the mechanism of RBC stiffening in a bloodstain is not completely known, it may be important to investigate not only the influence of external factors but also intrinsic RBC and bloodstain properties.

5. Conclusion

This chapter focused on the usefulness of AFM technology in forensic investigations. Important forensic examination subjects such as fingermarks, textile fibers, document forgery, gunshot and explosive residues and PSA have been discussed. Special attention was paid to age determination of bloodstains.

The forensic power of minimally invasive AFM is undoubtedly the ability to measure mechanical characteristics of trace materials such as elastic modulus, adhesion forces, energy dissipation and dielectric properties. Moreover, current AFM systems deliver synoptic mapping of these characteristics and FD curves for each pixel. Particularly, the ability to measure changes in mechanical properties of forensic traces over time makes this technology potentially interesting for a CSI. The increase of the YM of RBCs over time as measure for the age of a bloodstain is an excellent example of the forensic usefulness of FS. This information could provide valuable insights regarding the time of death of a victim or link a suspect to the scene at the time the crime was committed. Temporal forensic information can also be used to support or refute statements of victims, suspects and witnesses, especially when the defense states that the forensic evidence at hand is not related to the crime.

Less forensic value can be ascribed to AFM imaging. There are several reasons for this. The roughness of a surface on which a forensic trace has been deposited can hamper a proper examination of a height image. Although this problem could partially be solved by using phase imaging, many forensic trace characterizations require a chemical identity as well. Phase imaging will provide evidence only for the presence of different chemical substances but cannot identify them. However, AFMs equipped with optical microscopy and Raman spectroscopy or surface enhanced Raman spectroscopy may pave the way for AFM imaging as valuable tool for forensic examinations. Most promising in this respect is tip-enhanced Raman spectroscopy using a gold-coated atomic AFM tip-substrate system [80].

To grasp the forensic applicability of AFM in actual casework, additional research as well as laboratory and crime scene validation studies is required. The growing availability of fully automatic AFM systems that operate outside isolation cabinets brings a forensic AFM application closer.

Acknowledgements

The authors wish to thank Amu Hosseinzoi (BA.Sc.) for his tremendous commitment to a forensic application of AFM, particularly with respect to the RBC elasticity measurements.

Conflict of interest

There is no conflict of interest.

Thanks

Thanks to the great and close collaboration between Leiden Institute of Physics (LION, Leiden University, Leiden, The Netherlands) and the Netherlands Forensic Institute (the Hague, The Netherlands) we were able to write this chapter and perform our own forensic FS studies.

Author details

Threes Smijs[1,2]* and Federica Galli[2]

*Address all correspondence to: t.smijs@nfi.minvenj.nl

1 The Netherlands Forensic Institute, The Hague, The Netherlands

2 Leiden University, Leiden, The Netherlands

References

[1] Zou KN, Gui C, Gao Y, et al. Source identification of human biological materials and its prospect in forensic science. Fa Yi Xue Za Zhi. 2016;**32**:204-210. DOI: 10.1016/S1369-7021 (09)70167-1

[2] Almog J. Forensic science does not start in the lab: The concept of diagnostic field tests. Journal of Forensic Sciences. 2006;**51**:1228-1234. DOI: 10.1111/j.1556-4029.2006.00256.x

[3] Pitkethly M. Nanotechnology and forensics. Materials Today. 2009;**12**(6). DOI: 10.1016/ S1369-7021(09)70167-1

[4] Chen Y. Forensic applications of nanotechnology. Journal of the Chinese Chemical Society. 2012;**58**:828-835. DOI: 10.1002/jccs.201190129

[5] Inman K, Rudin N. Principles and Practice of Criminalistics—The Profession of Forensic Science. Boca Raton, Florida, USA: CRC Press; 2001

[6] Inman K, Rudin N. The origin of evidence. Forensic Science International. 2002;**126**:11-16. DOI: 10.1016/S0379-0738(02)00031-2

[7] Smijs T, Galli F, van Asten A. Forensic potential of atomic force microscopy. Forensic Chemistry. 2016;**2**:93-104. DOI: 10.1016/j.forc.2016.10.005

[8] Smijs T, Galli F. Forensic application of atomic force microscopy for age determination of bloodstains. Journal of Forensic Investigation. 2017;**5**:1-4

[9] JPK Instruments: Nano Wizard AFM handbook. version 2-2a; 2012. http://www.nanophys.kth.se/nanophys/facilities/nfl/afm/jpk/manuf-manuals/handbook-2.2a.pdf- [Accessed: March 17, 2016]

[10] Radmacher M, Cleveland JP, Fritz M, et al. Mapping interaction forces with the atomic force microscope. Biophysical Journal. 1994;**66**:2159-2165. DOI: 10.1016/S0006-3495(94)81011-2

[11] Sahin O, Magonov S, Su C, et al. An atomic force microscope tip designed to measure time-varying nanomechanical forces. Nature Nanotechnology. 2007;**2**:507-514. DOI: 10.1038/nnano.2007.226

[12] Rosa-Zeiser A, Weilandt E, Hild S, et al. The simultaneous measurement of elastic, electrostatic and adhesive properties by scanning force microscopy: Pulsed-force mode operation. Measurement Science and Technology. 1997;**8**:1333-1338. DOI: 10.1088/0957-0233/8/11/020

[13] Pittinger B, Erina N, Su C. Bruker Appl. Note. AN128; 2010. 1

[14] Rabe U, Janser K, Arnold W. Vibrations of free and surface-coupled atomic force microscope cantilevers: Theory and experiment. The Review of Scientific Instruments. 1996;**67**: 3281-3293. DOI: 10.1063/1.1147409

[15] Garcia R, Proksch R. Nanomechanical mapping of soft matter by bimodal force microscopy. European Polymer Journal. 2017;**49**:1897-1906

[16] Leite FL, Mattoso LHC, Oliveira ON, et al. The atomic force spectroscopy as a tool to investigate surface forces: Basic principles and applications. In: Méndez-Vilas A, Diaz J, editors. Modern Research and Educational Topics in Microscopy. FORMATEX; 2007. pp. 747-757

[17] Boularas A, Baudoin F, Villeneuve-Faure C, et al. Multi-dimensional modelling of electrostatic force distance curve over dielectric surface: Influence of tip geometry and correlation with experiment. Journal of Applied Physics. 2014;**116**. 084106–1–084106-11. DOI: doi.org/10.1063/1.4894147

[18] DelRio FW, Cook RF. Quantitative scanning probe microscopy for nanomechanical forensics. Experimental Mechanics. 2017;**57**:1045-1055. DOI: 10.1007/s11340-016-0238-y

[19] Durkan C, Wang N. Nanometre-scale investigations by atomic force microscopy into the effect of different treatments on the surface structure of hair. International Journal of Cosmetic Science. 2014;**36**:598-605. DOI: 10.1111/ics.12161

[20] Wada H, Usukura H, Sugawara M, et al. Relationship between the local stiffness of the outer hair cell along the cell axis and its ultrastructure observed by atomic force microscopy. Hearing Research. 2003;**177**:61-70. DOI: 10.1016/S0378-5955(02)00798-0

[21] Gurden SP, Monteiro VF, Longo E, et al. Quantitative analysis and classification of AFM images of human hair. Journal of Microscopy. 2004;**215**:13-23. DOI: 10.1111/j.0022-2720.2004.01350.x

[22] Tomes C, Jones JT, Carr CM, et al. Three-dimensional imaging and analysis of the surface of hair fibres using scanning electron microscopy. International Journal of Cosmetic Science. 2007;**29**:293-299. DOI: 10.1111/j.1467-2494.2007.00382.x

[23] Jeong KH, Kim KS, Lee GJ, et al. Investigation of aging effects in human hair using atomic force microscopy. Skin Research and Technology. 2011;**17**:63-68. DOI: 10.1111/j.1600-0846.2010.00466.x

[24] Kim KS, Lee J, Jung MH, et al. Characterization of human ovarian teratoma hair by using AFM, FT-IR, and Raman spectroscopy. Microscopy Research and Technique. 2011;**74**: 1121-1126. DOI: 10.1002/jemt.21003

[25] Kasas S, Khanmy-Vital A, Dietler G. Examination of line crossings by atomic force microscopy. Forensic Science International. 2001;**119**:290-298. DOI: 10.1016/S0379-0738(00)00458-8

[26] Ellen D. Scientific Examination of Documents Methods and Techniques. Boca Raton, USA: CRC Press; 2006

[27] Chen SZ, Tsai TL, Chen YF. Forensic application of atomic force microscopy questioned document. Journal of the Chinese Chemical Society. 2012;**59**:283-288. DOI: 10.1002/jccs.201100739

[28] Brandao JM, Almeida NSM, Dixini PVM, et al. Documentoscopy by atomic force microscopy (AFM) coupled with Raman microspectroscopy: Applications in banknote and driver license analyses. Analytical Methods. 2016;**8**:771-784. DOI: 10.1039/c5ay03128a

[29] Bruschweiler W, Grieve MC. A study on the random distribution of a red acrylic target fibre. Science and Justice. 1997;**37**:85-89. DOI: 10.1016/S1355-0306(97)72152-X

[30] Canetta E, Montiel K, Adya AK. Morphological changes in textile fibres exposed to environmental stresses: Atomic force microscopic examination. Forensic Science International. 2009;**191**:6-14. DOI: 10.1016/j.forsciint.2009.05.022

[31] Goddard AJ, Hillman AR, Bond JW. High resolution imaging of latent fingerprints by localized corrosion on brass surfaces. Journal of Forensic Sciences. 2010;**55**:58-65. DOI: 10.1111/j.1556-4029.2009.01217.x

[32] D'Uffizi M, Falso G, Ingo GM, et al. Microchemical and micromorphological features of gunshot residue observed by combined use of AFM, SA-XPS and SEM + EDS. Surface and Interface Analysis. 2016;**34**:502-506. DOI: 10.1002/sia.1348

[33] Oxley JC, Smith JL, Kirschenbaum LJ, et al. Accumulation of explosives in hair-Part 3: Binding site study. Journal of Forensic Sciences. 2012;**57**:623-635. DOI: 10.1111/j.1556-4029.2011.02020.x

[34] Poletti G, Orsini F, Lenardi C, et al. A comparative study between AFM and SEM imaging on human scalp hair. Journal of Microscopy. 2003;**211**:249-255. DOI: 10.1046/j.1365-2818.2003.01220.x

[35] Silva CS, Pimentel MF, Honorato RS, et al. Near infrared hyperspectral imaging for forensic analysis of document forgery. Analyst. 2014;**139**:5176-5184. DOI: 10.1039/c4an00961d

[36] Claybourn M, Ansell M. Using Raman spectroscopy to solve crime: Inks, questioned documents and fraud. Science and Justice. 2000;**40**:261-271. DOI: 10.1016/S1355-0306(00)71996-4

[37] Braz A, Lopez-Lopez M, Garcia-Ruiz C. Studying the variability in the Raman signature of writing pen inks. Forensic Science International. 2014;**245C**:38-44. DOI: 10.1016/j.forsciint.2014.10.014

[38] Meilero PP, García-Ruiz C. Spectroscopic techniques for the forensic analysis of textile fibers. Applied Spectroscopy Reviews. 2016;**51**:278-301. DOI: 10.1080/05704928.2015.1132720

[39] Koenig K, Girod A, Weyermann C. Identification of wax esters in fingermark residues by GC/MS and their potential use as aging parameters. Journal of Forensic Identification. 2011;**61**:652-676

[40] Bailey MJ, Bradshaw R, Francese S, et al. Rapid detection of cocaine, benzoylecgonine and methylecgonine in fingerprints using surface mass spectrometry. Analyst. 2015;**140**:6254-6259. DOI: 10.1039/c5an00112a

[41] Chan KL, Kazarian SG. Detection of trace materials with Fourier transform infrared spectroscopy using a multi-channel detector. Analyst. 2006;**131**:126-131. DOI: 10.1039/b511243e

[42] Dikshitulu YS, Prasad L, Pal JN, et al. Aging studies on fingerprint residues using thin-layer and high performance liquid chromatography. Forensic Science International. 1986;**31**:261-266. DOI: 10.1016/0379-0738(86)90165-9

[43] Connatser RM, Prokes SM, Glembocki OJ, et al. Toward surface-enhanced Raman imaging of latent fingerprints. Journal of Forensic Sciences. 2010;**55**:1462-1470. DOI: 10.1111/j.1556-4029.2010.01484.x

[44] Dutelle AW. Fingerprint evidence. In: An introduction to crime scene investigation. Burlington, USA: Jones and Bartlett Learning; 2014. pp. 169-202

[45] Bersellini C, Garofano L, Giannetto M, et al. Development of latent fingerprints on metallic surfaces using electropolymerization processes. Journal of Forensic Sciences. 2001;**46**:871-877. DOI: 10.1520/JFS15060J

[46] van Dam A, Aalders MC, Todorovski T, et al. On the autofluorescence of aged fingermarks. Forensic Science International. 2016;**258**:19-25. DOI: 10.1016/j.forsciint.2015.11.002

[47] Girod A, Ramotowski R, Lambrechts S, et al. Fingermark age determinations: Legal considerations, review of the literature and practical propositions. Forensic Science International. 2016;**262**:212-226. DOI: 10.1016/j.forsciint.2016.03.021

[48] Girod A, Xiao L, Reedy B, et al. Fingermark initial composition and aging using Fourier transform infrared microscopy (mu-FTIR). Forensic Science International. 2015;**254**:185-196. DOI: 10.1016/j.forsciint.2015.07.022

[49] Poppa P, Porta D, Gibelli D, et al. Detection of blunt, sharp force and gunshot lesions on burnt remains: A cautionary note. The American Journal of Forensic Medicine and Pathology. 2011;**32**:275-279. DOI: 10.1097/PAF.0b013e3182198761

[50] Aliste M, Chavez LG. Analysis of gunshot residues as trace in nasal mucus by GFAAS. Forensic Science International. 2016;**261**:14-18. DOI: 10.1016/j.forsciint.2016.01.034

[51] Aksoy C, Bora T, Senocak N, et al. A new method to reduce false positives due to antimony in detection of gunshot residues. Forensic Science International. 2015;**250**:87-90. DOI: 10.1016/j.forsciint.2015.03.006

[52] Abrego Z, Ugarte A, Unceta N, et al. Unambiguous characterization of gunshot residue particles using scanning laser ablation and inductively coupled plasma-mass spectrometry. Analytical Chemistry. 2012;**84**:2402-2409. DOI: 10.1021/ac203155r

[53] French J, Morgan R. An experimental investigation of the indirect transfer and deposition of gunshot residue: Further studies carried out with SEM-EDX analysis. Forensic Science International. 2015;**247**:14-17. DOI: 10.1016/j.forsciint.2014.10.023

[54] Taudte RV, Beavis A, Blanes L, et al. Detection of gunshot residues using mass spectrometry. BioMed Research International. 2014;**2014**:965403. DOI: 10.1155/2014/965403

[55] Coumbaros J, Kirkbride KP, Klass G, et al. Characterisation of 0.22 caliber rimfire gunshot residues by time-of-flight secondary ion mass spectrometry (TOF-SIMS): A preliminary study. Forensic Science International. 2001;**119**:72-81. DOI: 10.1016/S0379-0738(00)00421-7

[56] Abrego Z, Grijalba N, Unceta N, et al. A novel method for the identification of inorganic and organic gunshot residue particles of lead-free ammunitions from the hands of shooters using scanning laser ablation-ICPMS and Raman micro-spectroscopy. Analyst. 2014;**139**:6232-6241. DOI: 10.1039/c4an01051e

[57] Mou Y, Lakadwar J, Rabalais JW. Evaluation of shooting distance by AFM and FTIR/ATR analysis of GSR. Journal of Forensic Sciences. 2008;**53**:1381-1386. DOI: 10.1111/j.1556-4029.2008.00854.x

[58] Wu Y, Hu Y, Cai J, et al. Time-dependent surface adhesive force and morphology of RBC measured by AFM. Micron. 2009;**40**:359-364. DOI: 10.1016/j.micron.2008.10.003

[59] Strasser S, Zink A, Kada G, et al. Age determination of blood spots in forensic medicine by force spectroscopy. Forensic Science International. 2007;**170**:8-14. DOI: 10.1016/j.forsciint.2006.08.023

[60] Girasole M, Dinarelli S, Boumis G. Structure and function in native and pathological erythrocytes: A quantitative view from the nanoscale. Micron. 2012;**43**:1273-1286. DOI: 10.1016/j.micron.2012.03.019

[61] Chen Y, Cai J. Membrane deformation of unfixed erythrocytes in air with time lapse investigated by tapping mode atomic force microscopy. Micron. 2006;**37**:339-346. DOI: 10.1016/j.micron.2005.11.011

[62] Max E, Hafner W, Bartels FW, et al. A novel AFM based method for force measurements between individual hair strands. Ultramicroscopy. 2010;**110**:320-324. DOI: 10.1016/j.ultramic.2010.01.003

[63] Clifford CA, Sano N, Doyle P, et al. Nanomechanical measurements of hair as an example of micro-fibre analysis using atomic force microscopy nanoindentation. Ultramicroscopy. 2012;**114**:38-45. DOI: 10.1016/j.ultramic.2012.01.006

[64] Xu X, Mres J, Groven LJ, et al. Nanoscale characterization of mock explosive materials using advanced atomic force microscopy methods. Journal of Energetic Materials. 2015;**33**:51-65. DOI: 10.1080/07370652.2014.889780

[65] Paiva A, Sheller N, Foster MD. Study of the surface adhesion of pressure-sensitive adhesives by atomic force microscopy and spherical indenter tests. Macromolecules. 2000;**33**:1878-1881. DOI: 10.1021/ma990765v

[66] Canetta E, Adya AK. Atomic force microscopic investigation of commercial pressure sensitive adhesives for forensic analysis. Forensic Science International. 2011;**210**:16-25. DOI: 10.1016/j.forsciint.2011.01.029

[67] Paiva A, Sheller N, Foster MD. Microindentation and Nanoindentation studies of aging in pressure-sensitive adhesives. Macromolecules. 2001;**34**:2269-2276. DOI: 10.1021/ma0002343

[68] Stark M, Moller C, Muller DJ, et al. From images to interactions: High-resolution phase imaging in tapping-mode atomic force microscopy. Biophysical Journal. 2001;**80**:3009-3018. DOI: 10.1016/S0006-3495(01)76266-2

[69] Martinez NF, Garcia R. Measuring phase shifts and energy dissipation with amplitude modulation atomic force microscopy. Nanotechnology. 2006;**17**:S167-S172. DOI: 10.1088/0957-4484/17/7/S11

[70] Bunsell AR. Introduction to fibre tensile properties and failure. In: Handbook of Tensile Properties of Textile and Technical Fibres. Cambridge: Woodhead Publishing; 2009. pp. 1-17

[71] Katz E, Halamek J. Forensic Sciences: A Multidisciplinary Approach. Weinheim, Germany: Wiley-VCH; 2016

[72] Kumooka Y. Discrimination of rubber-based pressure sensitive adhesives by size exclusion chromatography. Forensic Science International. 2007;**171**:5-8. DOI: 10.1016/j.forsciint.2006.09.005

[73] Cui Y, Frank SG. Characterization of supersaturated lidocaine/polyacrylate pressure sensitive adhesive systems: Thermal analysis and FT-IR. Journal of Pharmaceutical Sciences. 2006;**95**:701-713. DOI: 10.1002/jps.20445

[74] Fang C, Lin ZX. Effect of propyleneimine external cross-linker on the properties of acrylate latex pressure sensitive adhesives. International Journal of Adhesion and Adhesives. 2015;**61**:1-7. DOI: 10.1016/j.ijadhadh.2015.04.009

[75] Sun ZW, Quan YK, Sun YY. Elemental analysis of white electrical tapes by wavelength dispersive X-ray fluorescence spectrometry. Forensic Science International. 2013;**232**:169-172. DOI: 10.1016/j.forsciint.2013.07.009

[76] Khalina M, Sanei M, Mobarakeh HS, et al. Preparation of acrylic/silica nanocomposites latexes with potential application in pressure sensitive adhesive. International Journal of Adhesion and Adhesives. 2015;**58**:21-27. DOI: 10.1016/j.ijadhadh.2014.12.007

[77] Edelman G, van Leeuwen TG, Aalders MC. Hyperspectral imaging for the age estimation of blood stains at the crime scene. Forensic Science International. 2012;**223**:72-77. DOI: 10.1016/j.forsciint.2012.08.003

[78] Sneddon IN. The relation between load and penetration in the Axi-symmetric Boussinesq problem for a punch of arbitrary profile. International Journal of Engineering Science. 1965;**3**:47-57. DOI: 10.1016/0020-7225(65)90019-4

[79] Bilodeau GG. Regular pyramid punch problem. Journal of Applied Mechanics. 1992;**59**: 519-523. DOI: 10.1115/1.2893754

[80] Meng L, Huang T, Wang X, et al. Gold-coated AFM tips for tip-enhanced Raman spectroscopy: Theoretical calculation and experimental demonstration. Optics Express. 2015;**23**: 13804-13813. DOI: 10.1364/OE.23.013804

Chapter 4

Scanning Probe Techniques for Characterization of Vertically Aligned Carbon Nanotubes

Marina V. Il'ina, Oleg I. Il'in, Vladimir A. Smirnov,
Yuriy F. Blinov, Boris G. Konoplev and Oleg A. Ageev

Additional information is available at the end of the chapter

http://dx.doi.org/10.5772/intechopen.78061

Abstract

This chapter presents the results of experimental studies of the electrical, mechanical and geometric parameters of vertically aligned carbon nanotubes (VA CNTs) using scanning probe microscopy (SPM). This chapter also presents the features and difficulties of characterization of VA CNTs in different scanning modes of the SPM. Advanced techniques for VA CNT characterization (the height, Young's modulus, resistivity, adhesion and piezoelectric response) taking into account the features of the SPM modes are described. The proposed techniques allow to overcome the difficulties associated with the vertical orientation and high aspect ratio of nanotubes in determining the electrical and mechanical parameters of the VA CNTs by standard methods. The results can be used in the development of diagnostic methods as well as in nanoelectronics and nanosystem devices based on vertically aligned carbon nanotubes (memory elements, adhesive structures, nanoelectromechanical switches, emission structures, etc.).

Keywords: nanotechnology, carbon nanotube, scanning probe microscopy, resistivity, Young's modulus, adhesion, piezoelectric response

1. Introduction

Precise parameters of vertically aligned carbon nanotubes (VA CNTs) control geometric parameters, resistivity, Young's modulus, adhesion, strength, and so on, and are a prerequisite for the devices of nanoelectronics and nanosystems creation on their basis with reproducible and stable characteristics [1–4]. However, the determination of these parameters in the carbon nanotubes (CNTs) by standard diagnostic methods is difficult due to the vertical orientation and the high aspect ratio of the nanotubes.

Thus, the application of traditional experimental methods for determining mechanical parameters (direct tensile load, pulsed dynamic method, etc.) is complicated due to the size of the VA CNT and also because of the need to fasten nanotubes on the substrate. In addition, as shown by the analysis of published data, the Young's modulus (one of the main mechanical parameters of CNTs) has a wide range of values in the range 0.4–6.85 TPa [5–13]. The values of the Young's modulus obtained experimentally are 2–3 times smaller [5, 7–10] calculated on the basis of theoretical models [11–13]. This may be due to the fact that the Young's modulus of CNTs depends essentially on the thickness of the CNT wall, which in practice is several times larger than values used in theoretical calculations [14].

The most widely used methods for investigating the electrical properties of microstructures require the formation of contact areas of several micrometers in size at the tops of the VA CNTs. This fact significantly limits the possibilities of using these methods to determine the electrical parameters of individual vertically aligned nanotubes because of their small transverse dimensions [15]. The work on quantitative evaluation and study of VA CNT adhesion to substrate is not numerous due to the complicated nature of interaction between the substrate, the catalytic center and the nanotube during the growth of the carbon nanotubes and also the need to manipulate individual nanotubes during experimental studies [16–18].

Thus, the tasks associated with development of new nanodiagnostic techniques to determine the geometric, mechanical and electrical parameters of VA CNTs are relevant in connection with the need to control and study parameters of individual vertically aligned carbon nanotubes, elements and devices on their basis and also in connection with requirements for development of metrological support for nanotechnologies.

A promising method for developing such nanodiagnostic techniques is a scanning probe microscopy (SPM) method [19–25]. This method does not require additional fixation of VA CNTs, special sample preparation and formation of contact areas on their tops. However, the determination of quantitative values of individual carbon nanotube parameters based on results obtained by SPM requires analysis of measurement process and development on its basis of techniques for determining parameters of CNTs taking into account the features of the SPM method.

This chapter describes unique techniques for determining the height, Young's modulus, bending stiffness, resistivity and adhesion to a substrate of vertically aligned carbon nanotubes, based on the methods of scanning probe microscopy. Described techniques can be used for nanodiagnostics of parameters of individual vertically aligned carbon nanotubes and for creation of nanoelectronic elements and devices on their basis [1, 11, 21, 23, 25–29].

2. Scanning probe techniques for characterization of vertically aligned carbon nanotubes

2.1. Experimental samples and equipment

The experimental samples of vertically aligned carbon nanotubes were grown using the NANOFAB NTC-9 nanotechnology multi-functional complex (NT-MDT, Russia) using the plasma-enhanced chemical vapor deposition (PECVD) method [2, 25, 30, 31]. A silicon wafer with a deposited titanium film and a nickel catalytic film with thicknesses of 20 and 10 nm, respectively, was used as the initial substrate. Acetylene was used as the reaction gas. The growth of VA CNT arrays was carried out at a pressure of 4.5 Torr and a temperature of 750°C. The growth parameters for experimental samples were characterized by the growth time, the acetylene feed rate and current flowing in the system. The structural analysis of the VA CNT arrays was conducted by the transmission electron microscopy (TEM) using Tecnai Osiris (FEI, Netherlands) and the Raman spectrometer Renishaw InVia Reflex (Renishaw plc, UK). Analysis of TEM images and Raman spectra showed that the experimental samples were multi-walled carbon nanotubes [3]. Surface investigations of the obtained VA CNT arrays were carried out using a scanning electron microscope (SEM) Nova Nanolab 600 (FEI Company, Netherlands).

Experimental studies of geometric, mechanical, electrical and adhesive properties of VA CNT arrays were carried out using the Ntegra probe nanolaboraty (PNL) (NT-MDT, Russia). To process the experimental data, the ImageAnalysis application package was used. A commercially available silicon cantilever of the NSG 20 brand was used as the probe of the atomic force microscope in developing the technique for determining the height of the VA CNT array. Investigations of the mechanical properties of carbon nanotube were carried out at Ntegra PNL by an integrated scanning hardness nanotester. The indenter was a diamond triangular pyramid of Berkovich with an angle at the vertex between the edge and height $\theta = 70°$. Investigations of the electrical parameters of VA CNTs were carried out by scanning tunneling microscopy (STM) method in spectroscopy mode with a distance between the STM probe and the VA CNT top equal to 0.5 nm. A tungsten probe made by an electrochemical method was used as a STM probe. Experimental studies of the VA CNT adhesion to a substrate were carried out by atomic-force microscopy (AFM) in current spectroscopy mode, with a distance between CNT top and an AFM probe equal to 1 nm. Experimental studies of piezoelectric response of VA CNT were performed by the AFM method in force spectroscopy mode. A commercially available silicon cantilever with a platinum coating of NSG11/Pt brand with a tip radius of 20 nm was used as an AFM probe.

2.2. A technique for determining the height of a vertically aligned carbon nanotube array

One of the promising methods for studying nanoscale structures is the atomic force microscopy, which allows one to determine the parameters of nanostructures without special sample preparation and to modify them by probe nanolithography methods [25]. The main difficulty in the study of VA CNT arrays by the AFM method is the mobility of nanotubes in their interaction with the probe. In addition, at a high density of carbon nanotubes in the array, the depth of penetration of the AFM probe between the individual nanotubes is limited by the parameters of the probe itself (the radius of curvature and aspect ratio of tip sides), which can lead to the display on AFM images of nonindividual nanotubes in the array and their bundles [25].

The study of influence of AFM scanning mode (contact, semi-contact and noncontact) on the quality of the image obtained on the surface of VA CNT array (with a diameter $D = 80$ nm, length $L = 2$ μm and density of nanotubes in the array $n = 30$ μm^{-2}) shows that the carbon nanotubes "detach" from substrate-surface VA CNTs in AFM contact mode (**Figure 1a**). As a result, the AFM contact mode cannot be used for nanodiagnostics and for determining the

VA CNT geometric parameters [25]. The study of the VA CNT array in the semi-contact mode shows that the individual VA CNTs are combined into bundles when the probe is exposed (**Figure 1b**). The main disadvantage of AFM images of VA CNT array obtained in a semi-contact mode is the presence of a number of scanning artifacts caused by the high mobility of nanotubes during mechanical contact with the probe, and as a consequence, the relatively low resolution of this AFM mode (**Figure 1b**).

In addition, a partial destruction of the VA CNT array is possible when scanning in a semi-contact mode with a pressing force of the AFM probe to surface more than 10 μN and scanning frequency more than 1 Hz (**Figure 2**).

The usage of the AFM noncontact mode at which the probe interacts with the array surface only due to van der Waals forces [32] made it possible to obtain AFM images of bundles of vertically aligned carbon nanotubes with a higher resolution, without explicit artifacts (**Figure 1c**). In the noncontact mode, the individual nanotubes were also combined into bundles with a diameter of about 300 nm (**Figure 1c**) under the action of van der Waals forces [25]. Statistical processing of AFM images showed that the maximum height of the bundle was 2.52 μm, the average height was 1.27 ± 0.35 μm and the density of individual VA CNT bundles in the array was about 1.68 μm^{-2} [25].

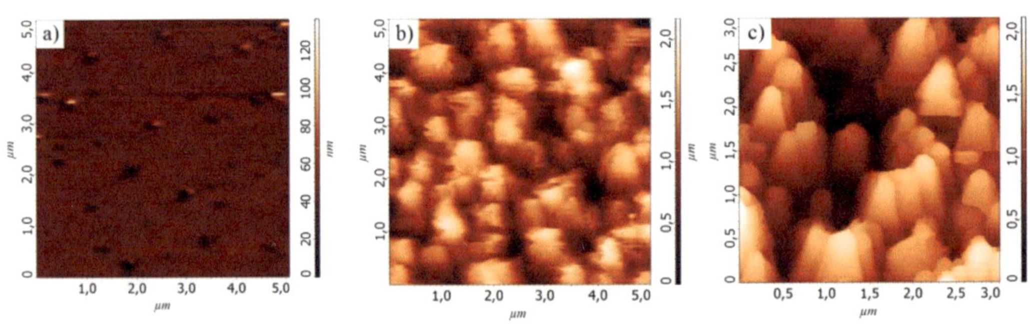

Figure 1. AFM images of the surface of the VA CNT array obtained by AFM: (a) in contact mode, (b) in semi-contact mode and (c) in noncontact mode [25].

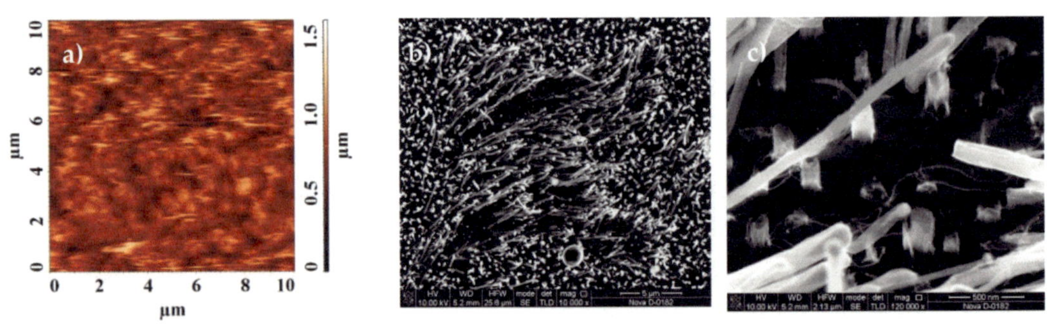

Figure 2. The images of the VA CNT array obtained after scanning in a semi-contact mode with a pressing force of the AFM probe to surface more than 10 μN: (a) AFM images; (b) and (c) SEM images.

Thus, it has been shown that the optimal scanning mode for determination of the geometric parameters of vertically aligned carbon nanotubes is the AFM noncontact mode, which allows one to obtain AFM images with a higher resolution resolving and without destroying the VA CNT structure. The results obtained by the AFM noncontact mode correlate well with the values of the geometrical parameters of the VA CNT determined by the SEM method.

Based on the obtained results, a technique for measuring a VA CNT array height was developed. This technique was based on sequential scanning in contact, then in semi-contact or noncontact modes of a VA CNT array with different areas [25]. So, the VA CNT array were scanned with an area of $10 \times 10\ \mu m^2$ and then $30 \times 30\ \mu m^2$ (**Figure 3a, b**). The analysis of the profilogram of the scanning area allowed to determine the maximum height of the VA CNT array equal to 1.98 μm and the average height of VA CNTs in the array equal to 1.12 ± 0.45 μm (**Figure 3c**).

The developed technique for measuring the VA CNT array height allows to determine the height value with a higher reliability than a semi-contact or noncontact mode in that the measurement of this parameter is made relative to substrate's surface and not the greatest penetration depth of the AFM probe into the array. Moreover, the developed technique makes it possible to obtain and automatically process a statistical set of the geometric parameters' values of carbon nanotubes in contrast to the SEM method.

2.3. A technique for determining the Young's modulus and bending stiffness of vertically aligned carbon nanotubes

One of the promising methods for determining the Young's modulus of vertically aligned CNTs is the nanoindentation method based on indenting a solid needle (indenter) into the array by applying an external load and obtaining the dependence of the penetration depth of the indenter into the array from nanoindentation force [11, 22]. A schematic process of nanoindentation of a vertically aligned carbon nanotubes array is shown in **Figure 4**. Initially, the indenter is in the approached state, then the load is applied and the indenter interacts with the array surface and touches the first VA CNT at the depth h_1; with a further increase in the load, the first nanotube begins to bend and the indenter touches the second tube at the depth h_2 (**Figure 4a**). At a given depth h, the indenter interacts with the i-number of VA CNTs, each of which is deflected by a distance w_0 (**Figure 4b**), depending on the initial touch depth of the indenter with the i-tube and the geometry of the indenter [13]:

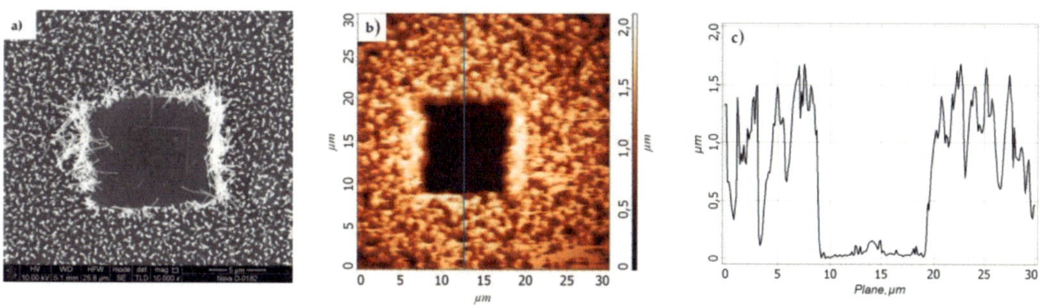

Figure 3. Measuring a VA CNT array height based on developed technique: (a) SEM image, (b) AFM image and (c) profilogram along the line [25].

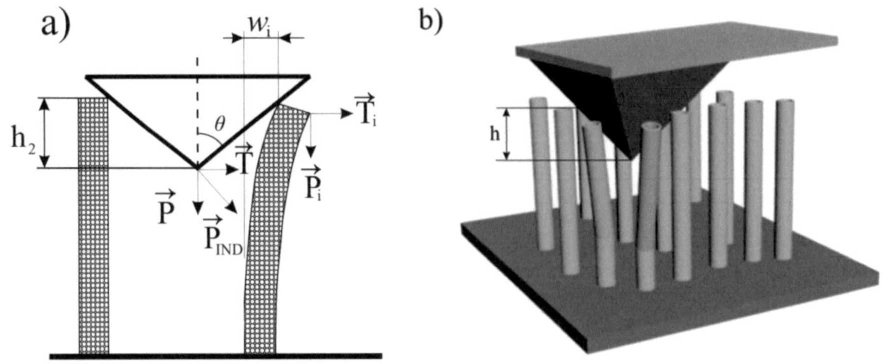

Figure 4. Schematics of the nanoindentation process of the VA CNT array: (a) the deflection of the first nanotube and the touching of the second nanotube of the indenter with increasing applied force; (b) the interaction of the indenter with *i* nanotubes at depth *h* [29].

$$w_0 = (h - h_i)\text{tg}\theta. \tag{1}$$

On the other hand, the nanoindentation process can be considered using a micromechanical model. This model is based on the beam theory, according to which an individual vertically aligned nanotube is an elastic hollow cylindrical rod fixed at one end. In this case, the elastic deflection w_0 of a carbon nanotube with an outer diameter D and a height L when an external load is applied is described by the equation [11]:

$$w_0 = \frac{T}{P}\left(\frac{\text{tg}kL}{k} - L\right), \tag{2}$$

where P and T are the forces acting on the nanotube parallel and perpendicular to its axis, respectively (**Figure 4a**); $k = (P/(YI)_{eff})^{1/2}$, $(EI)_{eff}$ is the effective bending stiffness of a VA CNT with an effective moment of inertia equal to $I_{eff} \approx \pi D^4/64$ [11].

Earlier, a method was developed for determining the mechanical parameters of a VA CNT based on the nanoindentation using a micromechanical model [11]. The main shortcoming of this method was the use of bending stiffness as a rigidity parameter of theoretical and experimental dependencies, from which the value of the Young's modulus of the VA CNT is calculated [11] This fact significantly reduced the reliability of the obtained results. To eliminate this shortcoming, we proposed a technique for determining the bending stiffness $(EI)_{eff}$ for each i-tube interacting with the indenter using experimental dependences [22, 28, 29]. For this the force P acting on one CNT was represented as [22]:

$$P = (P_{IND} - P_i)\cos\theta / i, \tag{3}$$

where P_{IND} is the indentation force representing the vector sum of the forces P and T (**Figure 4a**), P_i is the indentation force corresponding to the depth hi of touching of the i-tube and i is the number of nanotubes interacting with the indenter at the P_{IND}. The values of the P_{IND} and P_i forces are determined from the experimental dependences obtained in the nanoindentation process of the VA CNT array.

Thus, Young's modulus is determined from the following expression based on the technique developed by us [22]:

$$Y = \frac{64(P_{IND} - P_i)\cos\theta}{\pi\, iD^2 k^2}. \tag{4}$$

The number of nanotubes i interacting with the indenter for a given indentation force is defined as the product of the total interaction area of the indenter S at penetration depth h and density of the VA CNTs in an array. The total area S is the sum of the cross-section area of the indenter S_{ind} at depth h, the interaction area by the perimeter of indenter S_{per} and the cross-sectional area of the nanotube S_{CNT}. Due to the fact that the Berkovich indenter is a regular triangular pyramid with the angle at the vertex $\theta = 70°$, the height of the triangle lying at the base of the pyramid is $a = 1.5\ h{\cdot}tg70°$ and the value of the edge of the base is $c = 2a{\cdot}tg30°$. The interaction area of the indenter S at the penetration depth h with one tube is [22]:

$$S = i/n = S_{ind} + S_{per} + S_{CNT} = 9.853\,h^2 + 14.355\,hD_0 + 3.141\,D_0^2/4. \tag{5}$$

This dependence allows us to determine not only the number of tubes interacting with the indenter at a depth h but also the depth h_i at which the indenter touches the i-tube.

The technique proposed by us technique makes it possible to calculate the Young's modulus of individual VA CNT directly from a set of the experimental dependences obtained in the nanoindentation process [22] The limits of applicability of the technique are determined by the aspect ratio of vertically aligned carbon and the nanotube deflection value. The maximum value of the deflection of CNTs in the nanoindentation process, in which the interaction of the indenter with the VA CNT array is still described by the beam theory, depends on the length of the carbon nanotube and is determined by the following expression [22]:

$$w_{max} = \left(0.2L - h_1\right)\mathrm{tg}\theta, \tag{6}$$

where $0.2\,L$ is the maximum penetration depth determined from the experimental dependences.

For the approbation of the developed technique, three experimental samples of VA CNT arrays with various geometric parameters were studied (**Figure 5**). Analysis of SEM images made it possible to estimate the diameter and height of carbon nanotubes, as well as the density of nanotubes in the array. The values of these parameters are presented in **Table 1**.

Using the obtained values of the geometric parameters of carbon nanotubes, their mechanical properties were investigated by the developed technique. The maximum value of an indentation force was 100 μN. The nanoindentation was carried out at six different points separated from each other by a distance of about 5 μm for each VA CNT array. **Figure 6** shows the experimental dependences obtained for three arrays of VA CNT.

Analysis of the dependences showed that the curve of the penetration depth of the indenter into the array on the indentation force is nonlinear. Two sections of the curve can be distinguished: the elastic (from 0 to 250 nm for the first mass, from 0 to 175 nm for the second mass, and from 0 to 250 nm for the third one) and inelastic (from 250 to 330 nm for the first array, from 175 to 275 nm for the second one and from 250 to 600 nm for the third one) interaction. Therefore, only the first section of the curves was used to calculate the Young's modulus of

Figure 5. SEM images of the investigated arrays of vertically aligned carbon nanotubes: (a) VA CNTs No 1, (b) VA CNTs No 2 and (c) VA CNTs No 3.

VA CNTs as the beam theory describes only small elastic deflections. Based on this, the total penetration depth of the indenter into the first array h was 250 nm, into the second one was 175 nm and into the third one was 250 nm. The corresponding indentation forces of P_{IND} were about 8, 10 and 3 μN, respectively (**Figure 6**).

Using the developed technique, the bending stiffness and Young's modulus were determined for each i-nanotube of the VA CNT interacting with the indenter at a depth h and the average values of these parameters were obtained for each VA CNT arrays. The results are shown in **Table 2**. The obtained values of the mechanical parameters of the VA CNT correlate well with the literature data [5–7, 9, 11, 22, 28].

An analysis of the results showed that the Young's modulus of a VA CNT increases with the increase in its length. This fact can be associated with decreasing the structural defects of the nanotube with increasing length. Analysis of the bending stiffness values showed that the value of this parameter increases with increasing diameter of the nanotubes. This is probably due to an increase in the number of inner layers in multi-walled carbon nanotubes and an increase in Van der Waals forces between layers, which leads to an additional resistance of the carbon nanotube to bending deformations during indentation [14].

Thus, the developed technique can be successfully used to measure the mechanical parameters of vertically aligned carbon nanotubes by the nanoindentation, as well as to study the effect of the geometric parameters of VA CNT on their Young's modulus [22, 28]. The developed technique can be used to determine the mechanical properties of nanotubes and nanowires from other materials [21].

Parameter	VA CNTs No 1	VA CNTs No 2	VA CNTs No 3
Height L, nm	650	1210	930
Diameter D, nm	44	36	51
Aspect ratio H	15	34	13
The VA CNTs density in array n, μm^{-2}	82	72	69

Table 1. Geometric parameters of the VA CNTs determined by the SEM method.

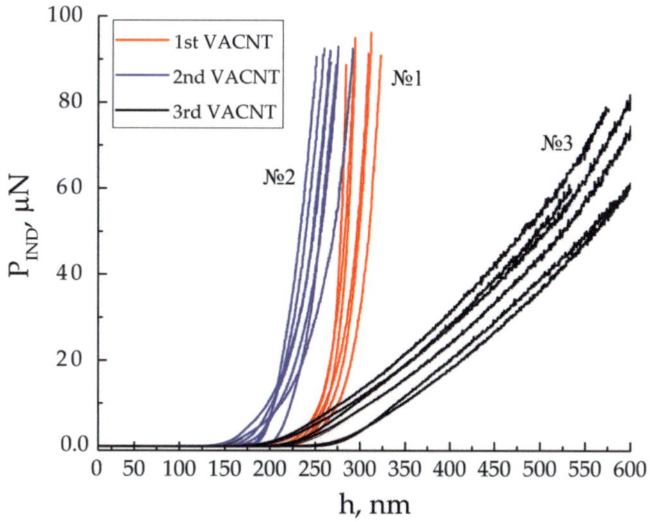

Figure 6. Experimental dependences obtained by the nanoindentation of the VA CNT arrays.

2.4. A technique for determining the resistivity of vertically aligned carbon nanotubes

Scanning probe microscopy is a precision method for studying the electrical properties of horizontal carbon nanotubes [33–35]. However, the study of vertically aligned carbon nanotube has difficulties due to the mobility of nanotubes upon contact with the SPM probe and the formation of VA CNT bundles on the application of an electric field [36]. In addition, the determination of the VA CNT resistivity based on the current–voltage characteristics (CVC) obtained by the SPM method requires an analysis of the measuring process the CVCs and the development of a technique taking into account the features of the SPM [23]. Earlier, it was shown that the value of the VA CNT resistivity obtained on the basis of CVCs obtained by the AFM method is higher [23] than the value presented in the literature [37]. This is due to the influence of contact of the AFM probe on the VA CNT top and the appearance of additional resistance in the measuring system. In addition, as a result of preliminary scanning of the VA CNT array in an AFM semi-contact mode, nanotubes form bundles, which makes it difficult to localize the probe over an individual VA CNT top and to study its electrical properties. The determination of the resistance of an individual VA CNT by the scanning tunneling microscopy (STM) method allows to overcome these difficulties because the resistance of the tunnel contact of the STM probe with the VA CNT top at a voltage of

Mechanical parameters	VA CNTs No 1	VA CNTs No 2	VA CNTs No 3
Bending stiffness, N·nm^2	0.112 ± 0.004	0.106 ± 0.005	0.195 ± 0.007
Young's modulus, TPa	1.15 ± 0.05	1.29 ± 0.08	0.59 ± 0.12

Table 2. The geometrical and mechanical parameters of the nanotubes in the VA CNT array determined by the AFM and the nanoindentation.

more than 1 V becomes insignificant and does not significantly affect the overall resistance of the system "STM probe/VA CNT/conductive layer/VA CNT array/contact." In addition, the formation of VA CNT bundles does not occur during the preliminary scanning by the STM method [23].

The schematic and the equivalent circuit of the measuring system of the VA CNT by the STMs method are shown in **Figure 7**. According to **Figure 7b**, the total resistance of the system R_{tot}^{STM} is [23]:

$$R_{tot}^{STM} = R_0^{STM} + R_{CNT/sub} + R_{CNT} + R_{tun}, \tag{7}$$

$$R_0^{STM} = R_{Me} + R_{CNTs/Me} + R_{CNTs} + R_{CNTs/sub} + R_{sub} + R_p^{STM}, \tag{8}$$

where R_0^{STM} is the total resistance of the conductive layer R_{sub}, the contact materials R_{Me} and the STM probe R_p^{STM}, as well as the array of nanotubes under contact $R_{CNTs/Me} + R_{CNTs} + R_{CNTs/sub}$, determined by STM spectroscopy of the array area without the VA CNTs (**Figure 7c**, **d**); R_{tun}—resistance of the tunnel contact between the STM probe and VA CNT; $R_{CNT} + R_{CNT/sub}$ is the total resistance of an individual vertically aligned carbon nanotube and the contact between the VA CNT and the conductive layer [23].

Earlier, it was shown that the contribution of the resistance of the tunnel contact decreases with increasing electric field strength and at large values of the strength one can take $R_{tun} \sim 0$ [38], where the total resistance of the individual VA CNT and the contact to the conductive layer can be taken as [23]:

$$R_{CNT}^{STM} = R_{tot}^{STM} - R_0^{STM}. \tag{9}$$

The resistance R_{tot}^{STM} was determined on the basis of the CVC obtained by STM spectroscopy on an individual VA CNT. The resistance R_0^{STM} was determined on the basis of the equivalent circuit for $R_{tun} \sim 0$ (**Figure 7d**) and the CVC obtained at the modified area of the VA CNT array. The resistivity of an individual VA CNT was defined as $\rho_{el} = R_{CNT}^{STM} \cdot S/L$.

The experimental approbation of the proposed technique was carried out on an experimental sample of VA CNT array (D = 70–120 nm, L = 2.2 μm, n = 8 μm^{-2}). The distance between the

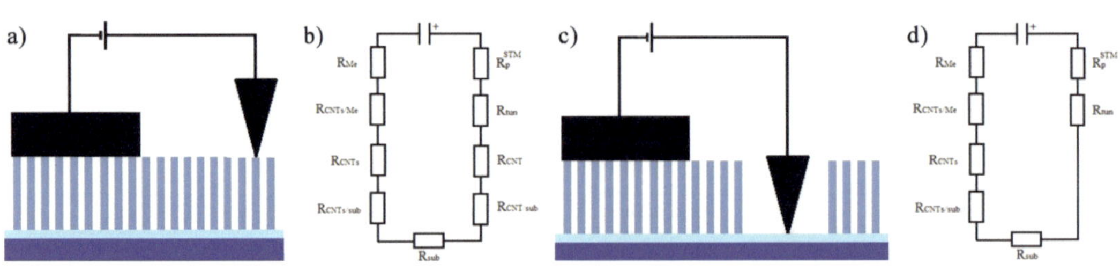

Figure 7. Measurement of the electrical parameters of the VA CNT by the STM method: (a, b) a schematic and equivalent circuit of measuring the total resistance; (c, d) a schematic and equivalent circuit without resistance of the VA CNT and its contacts [23].

STM probe and the VA CNT top was 0.5 nm. To localize the probe at the VA CNT top, a preliminary scanning of the array surface was performed using the STM method. The resulting STM image and the CVC of the VA CNT (solid line) are shown in **Figure 8**. The resistance of the "STM probe/conductive layer/VA CNT array/contact" system (**Figure 7c**) was measured additionally to exclude the resistance of the STM probe, the conductive layer, the contact material and the nanotube array below it from the total resistance system. To measure this resistance, the VA CNT array was previously scanned in the AFM contact mode using the technique presented in Section 2.2. The CVC of the modified area is shown in **Figure 8b** (dashed line).

Analysis of the obtained STM image of the VA CNT array showed that individual nanotubes are not combined into bundles due to the low density of VA CNTs in the array (**Figure 8a**). It allows investigating the electrical properties of individual VA CNTs. The diameter of the VA CNT is 118 ± 39 nm. The height of the VA CNT is not displayed correctly on the STM image due to the peculiarities of measuring the VA CNT array by the STM [36]. Based on the CVC of the individual VA CNT (**Figure 8b**), it can be concluded that the individual VA CNT exhibits two conduction states: high resistance when the voltage is varied from 0 to 10 V and low resistance when the voltage varies from 10 to 0 V, which is due to the manifestation of a memristor effect in VA CNT [3, 26, 29, 39, 40]. The low-resistance state of the VA CNT was used to determine the resistance of the nanotube, since there is no additional resistance in VA CNT associated with the internal electric field in the nanotube [3, 40].

The resistance of R_{tot}^{STM} was determined on the basis of the CVC, an individual VA CNT in low-resistance state, and was 108 kΩ (**Figure 8b**). The resistance R_0^{STM} was determined on the basis of the CVC obtained at the modified area of the VA CNT array (**Figure 8b**) and was 41 kΩ. From there, the total resistance of the individual VA CNT and the contact to the conductive layer R_{CNT}^{STM} was 67 kΩ. It was previously shown that a transition electrical resistivity of the contact of the VA CNT with the conducting layer is about 118.6 kΩ nm² ($1.186 \cdot 10^{-9}$ Ω cm²) [23]. Therefore, the resistance of the contact for the VA CNTs under study changes in the range $R_{CNTs/sub} = 4.1–12.8$ Ω. Thus, we have $R_{CNTs/sub} << R_{CNT}^{STM}$ and this resistance weakly contributes to the resistance of VA CNTs determined by the STM technique. Taking into account the geometric parameters of VA

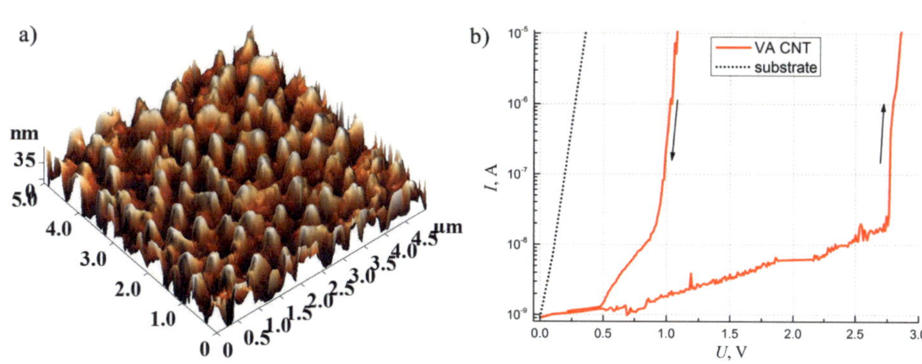

Figure 8. Investigation of the VA CNT array by the STM method: (a) 3D STM image of the VA CNT array; and (b) CVCs of an individual nanotube (solid line) and substrate (dashed line) [23].

CNT, the resistivity of VA CNT was (8.32 ± 3.18)·10^{-4} Ωm. The obtained value of the resistivity of multi-walled VA CNT correlates well with the literature data [41].

2.5. A technique for determining the adhesion of vertically aligned carbon nanotube to a substrate

The adhesion of VA CNT to a substrate W_a quantitatively characterized by the work of breaking the adhesive bond per unit area was assumed to be equal to the work of the internal elastic forces arising at the base of the VA CNT at the moment of detachment from substrate [42]:

$$W_a = \sigma \frac{\Delta L_0}{x_0} x_0 = \sigma \Delta L_0, \tag{10}$$

where $\sigma = k\Delta L_0/S$—mechanical stress arising in a VA CNT as a result of elongation; ΔL_0 is a VA CNT elongation at the point $x_0 = 0.01\ L$ at $t = t_0$.

From a practical point of view, the adhesion strength and the corresponding detachment force (the maximum force that can be applied along the axis of the nanotube without detaching it from the substrate) are of great importance. These parameters allow optimizing the design and operating parameters of emitters and memory elements based on VA CNT to prevent a VA CNT detaching from the substrate. The adhesion strength f_0 was defined as the force required for breaking the adhesive bond of the VA CNT with the substrate referred to the area of the adhesive contact. The value of adhesive strength is sensitive to the determination method and depends on the type of external load causing destruction of the adhesive contact. When applying a mechanical load to the VA CNT, the external force is expended on the deformation of VA CNT (F_x) and the break of the adhesion bond with the substrate ($F_a = W_a \cdot S/\Delta L_0$), and the adhesion strength of the VA CNT bond to the substrate is determined as [42]:

$$f_0 = (F_x(L, t_0) + F_a)/S = F_0/S. \tag{11}$$

When applying an external electric field, the adhesive strength $f_0^{'}$ is higher than adhesion strength f_0 by applying a mechanical load due to the attraction force $F_{at}(t)$ arising under the action of an external electric field which is expended not only on the deformation of VA CNT and the break of adhesive bond with the substrate but also on the polarization and creation of the conduction current VA CNT. The adhesive strength $f_0^{'}$ is determined as [42]:

$$f_0 = F_{at}(t_0)/S. \tag{12}$$

Thus, the value of the maximum force that can be applied to the "VA CNT/substrate" system without its destruction will be different for devices functioning in the field of mechanical loads of high electric field.

Experimental studies of adhesion were carried out on a VA CNT array (D = 70—120 nm, L = 2.2 μm and n = 8 μm^{-2}) when voltage pulses U were applied with an amplitude of 10 V to 30 V and duration of 1 s. The detachment of the VA CNT from the substrate was fixed by the absence of current on the reverse branch of the CVC (with the voltage from the maximum value to zero) obtained in the "substrate/VA CNT/AFM probe" system in the AFM spectroscopy mode (**Figure 9**). The nanotube is retained on the surface of the AFM probe by the van der Waals forces

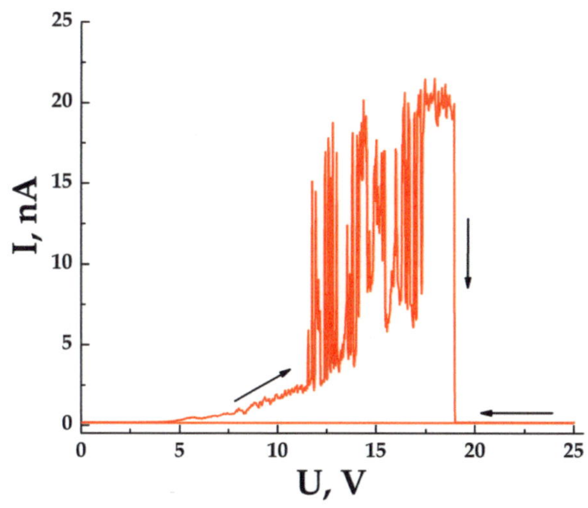

Figure 9. A CVC of the VA CNT during detachment from the substrate by AFM current spectroscopy.

after removal of the external electric field [19]. This effect was used to create critical-dimension atomic-force microscopy (CD-AFM) probes by depositing a carbon nanotube from a VA CNT array on the tip [19]. As shown by experimental studies, the detachment of a vertically aligned carbon nanotube from the substrate did not occur at U = 10 V; a single nanotube was detached with a reproducibility of about 30% at U = 15 V; a constant detachment of one or two VA CNTs was recorded at U = 20 V; several nanotubes were detached at U > 20 V. The voltage at which the detachment of a single VA CNT from the substrate was observed was likely to vary with the diameter of the nanotube. It should be noted that only van der Waals forces (~ 15 nN [3]).

Further, a modeling of the VA CNT deformation under the action of an external electric field was carried out to determine the quantitative values of the adhesion of VA CNTs to the substrate [3, 29, 40]. The results of the modeling showed that the VA CNT deformation increases with increasing voltage from 10 V to 30 V. So the elongation at the VA CNT base (x = 0.01 L) ΔL_0 is 0.037, 0.0391, 0.0404, 0.0419, 0.0441 and 0.0445 nm for a VA CNT with $D = 70$, 80, 90, 100, 110 and 120 nm, respectively, at corresponding voltage at the detachment (**Figure 10**). The elongation at the VA CNT top ($x = L$) ΔL_{max} is 1.25, 1.31, 1.36, 1.41, 1.48 and 1.49 nm, respectively (**Figure 10**). An analysis of the dependence of the elongation of a vertically aligned carbon nanotube (at $x = L$) on its diameter showed that at a voltage of 15 V ΔL_{max} of the VA CNT with a diameter of 70 nm is 1.24 nm and the ΔL_{max} increases to 1.49 nm with an increase in the diameter of VA CNT to 120 nm. Thus, voltage at the detachment of a VA CNT with $D = 80$ nm is equal to 16.3 V; with $D = 90$ nm, $U = 17.5$ V; with $D = 100$ nm, $U = 18.7$ V; with $D = 110$ nm, $U = 19.6$ V and; with $D = 120$ nm, $U = 20$ V (**Figure 10**) [42].

Taking into account the results of modeling, the adhesion of VA CNT ($D = 70-120$ nm) W_a increases from 0.55 to 1.19 mJ/m^2 with the increase in diameter according to Eq. (10). This fact probably connected with the increase in the number of carbon atoms interacting with the substrate per unit area of the adhesive contact, when the diameter of the VA CNT is increased and as a result of the increase in the layers of the multi-walled carbon nanotube. The corresponding adhesion force F_a of the individual VA CNT with increasing diameter varied from 92.5 nN to 226.1 nN. According to Eq. (11), the adhesive strength of the VA CNT to the substrate f_0 is 714.1 ± 138.4 MPa when a

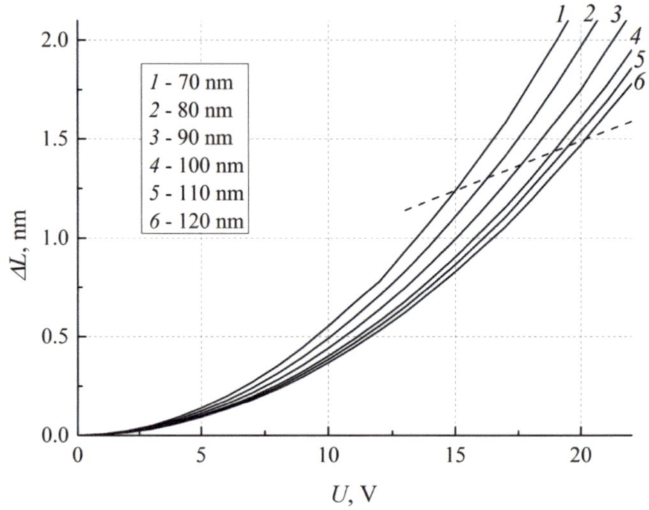

Figure 10. Dependence of the VA CNT elongation with a diameter of 70–120 nm (at $x = L$) on the voltage at time $t = t_0$. The dotted line indicates an elongation at which the VA CNT can be detached from the substrate [42].

mechanical load is applied. The corresponding detachment force F_0 varies from 1.93 to 10.33 μN depending on the diameter of the VA CNT (**Figure 11**, dependence 1).

The detachment force F_0 exceeds the adhesion force F_a by tens of times due to the considerable internal elastic forces arising in VA CNT under tension. According to Eq. (12), the adhesion strength of the VA CNT to the substrate f_0' is 1.43 ± 0.29 GPa when applying the external electric field. The corresponding detachment force is equal to the attractive force $F_{at}(t_0)$ at the moment of nanotube detachment from the substrate (varies from 3.83 μN to 20.02 μN for VA CNT with a diameter of 70–120 nm) (**Figure 11**, the dependence of 2) [42].

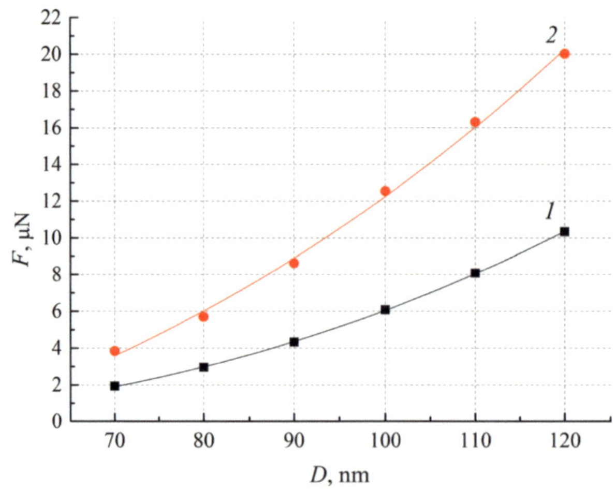

Figure 11. Dependence of the maximum force that can be applied along the axis of VA CNT without its detachment from the substrate, on the diameter of the nanotube: 1—When applying a mechanical load; and 2—When applying an external electric field. The points correspond to the values calculated on the basis of experimental and theoretical data; the solid line is the approximation of the obtained values [42].

Thus, the adhesion strength of a VA CNT to the substrate when applying an external electric field $f_0^{'}$ exceeds by almost two times the adhesion strength when applying mechanical load f_0. It allows using a VA CNT with diameter from 70 to 120 nm grown by PECVD in the development and creation of emitters and memory elements with an electric field strength up to 10^{10} V/m. The calculated adhesion force and adhesion strength are well correlated with available literature data [17, 18], which confirms the reliability of the results obtained.

2.6. A technique for determining the piezoelectric response of vertically aligned carbon nanotubes

Recent work in the field of investigation of electromechanical properties of carbon nanostructures has shown the possibility of manifesting flexoelectric and piezoelectric effects in them [3, 43–46]. In this connection, an urgent task was the development of a technique for determining the piezoelectric response of vertically aligned carbon nanotubes because the use of the piezoresponse force microscopy in this case is difficult.

To study the piezoelectric response of VA CNT, we proposed a technique based on AFM power spectroscopy with parallel detection of a current flowing in the "lower electrode/VA CNT/AFM probe" system using an AFM oscilloscope. As a result, deformation of VA CNTs was formed in the AFM force spectroscopy mode by mechanical pressing of the probe to its top and the current value generated by VA CNT was detected at a known mechanical load. Schematic of the measurement process is shown in **Figure 12a**.

The results of study of the VA CNT array (D = 34 ± 3 nm, L = 370 ± 40 nm and n = 47 μm^{-2}) on the basis of the proposed technique showed that the current in the "lower electrode/VA CNT/AFM probe" system did not flow when an AFM probe is approached to an individual nanotube; with increasing the pressing force of the probe from 0 to 0.5 μN, a current appeared the value of which varied from 0 to −24 nA, respectively; the current value decreased back to zero at the subsequent removal of the pressing force (**Figure 12b**).

To exclude the influence of the measurement system and the substrate on the measurement results, it is also necessary to carry out similar measurements on a substrate mechanically purified from VA CNT using the AFM contact mode. For this sample, the measurements of

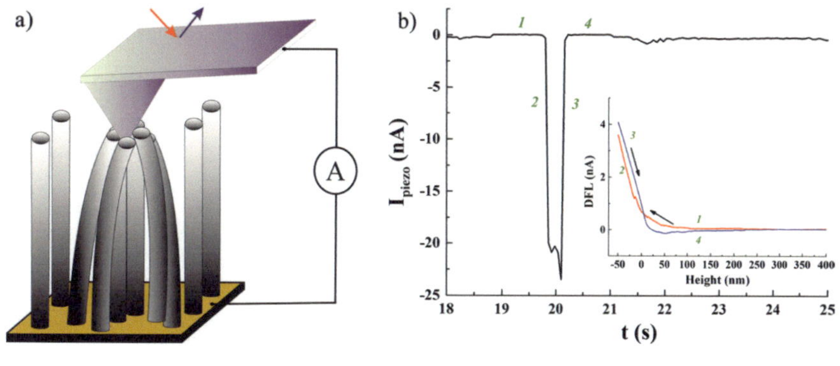

Figure 12. Study of the piezoelectric response of the VA CNT: (a) schematic of the measurement process; and (b) current-time dependence of a VA CNT and the corresponding AFM force spectroscopy [46].

the substrate showed the absence of a piezoelectric response when it deformed. The current flowing in the measurement system was constant and amounted to about 60 pA, which can be attributed to the systematic error of the PNL Ntegra measurement system.

Thus, the developed technique allows us to experimentally estimate the value of a current generated by deforming a carbon nanotube as a result of a direct piezoelectric effect. The obtained results correlate with the experimental and theoretical studies carried out by us earlier [3, 44].

3. Conclusion

Thus, unique techniques for nanodiagnostics of geometric, mechanical, electrical and adhesion properties of vertically aligned carbon nanotubes have been developed. The developed techniques were used for experimental investigations of the VA CNT. The obtained values of Young's modulus, bending stiffness, resistivity and adhesion of VA CNTs correlate well with the published data [5–7, 9, 11, 17, 18, 22, 28, 41], which confirm the reliability of the developed techniques. The developed techniques do not require additional sample preparation and can be used as express techniques for quality control of grown VA CNTs and elements of nanoelectronics and nanosystems based on them. The obtained results can be used for the development of nanodiagnostic methods, as well as for the design and fabrication of resistive energy-efficient memory elements with a high density of cells, adhesive structures, nanoelectromechanical switches and emission structures based on vertically aligned carbon nanotubes. Application of the developed techniques at the stage of interoperational control of the technological process of manufacturing such devices will increase reproducibility of their parameters and stability of operation.

The results were obtained using the equipment of Research and Education Center and the Center for collective use "Nanotechnologies" of Southern Federal University.

Acknowledgements

This work was financially supported by Russian Foundation for Basic Research (project No. 16-29-14023 ofi_m) and Internal grant of the Southern Federal University (project No. VnGr-07/2017-2026).

Conflict of interest

The authors declare no conflict of interest.

Author details

Marina V. Il'ina[1]*, Oleg I. Il'in[1], Vladimir A. Smirnov[1], Yuriy F. Blinov[1], Boris G. Konoplev[2] and Oleg A. Ageev[2]

*Address all correspondence to: mailina@sfedu.ru

1 Southern Federal University, Institute of Nanotechnologies, Electronics and Electronic Equipment Engineering, Taganrog, Russian Federation

2 Southern Federal University, Research and Education Center "Nanotechnologies", Taganrog, Russian Federation

References

[1] Choi WB, Bae E, Kang D, et al. Aligned carbon nanotubes for nanoelectronics. Nanotechnology. 2004;**15**:S512-S516. DOI: 10.1088/0957-4484/15/10/003

[2] Chen H, Roy A, Baek JB, Zhu L, Qu J, Dai L. Controlled growth and modification of vertically-aligned carbon nanotubes for multifunctional applications. Mater Sci Eng R Reports. 2010;**70**(3-6):63-91. DOI: 10.1016/j.mser.2010.06.003

[3] Il'ina MV, Il'in OI, Blinov YF, et al. Memristive switching mechanism of vertically aligned carbon nanotubes. Carbon. 2017;**123**:514-524. DOI: 10.1016/j.carbon.2017.07.090

[4] Lefebvre J, Antonov RD, Radosavljevic M, Lynch JF, Llaguno M, Johnson AT. Single-wall carbon nanotube based devices. Carbon. 2000;**38**(11-12):1745-1749. DOI: 10.1016/S0008-6223(00)00050-6

[5] Lourie O, Wagner HD. Evaluation of Young's modulus of carbon nanotubes by micro-Raman spectroscopy. Journal of Materials Research. 1998;**13**(9):2418-2422. DOI: 10.1557/JMR.1998.0336

[6] Ding W, Calabri L, Kohlhaas KM, Chen X, Dikin DA, Ruoff RS. Modulus, fracture strength, and brittle vs. plastic response of the outer shell of arc-grown multi-walled carbon nanotubes. Experimental Mechanics. 2007;**47**(1):25-36. DOI: 10.1007/s11340-006-9344-6

[7] Wong EW, Wong EW, Sheehan PE, Lieber CM. Nanobeam mechanics: Elasticity, strength, and toughness of nanorods and nanotubes. Science. 1997;**277**(5334):1971-1975. DOI: 10.1126/science.277.5334.1971

[8] Salvetat J-P, Kulik AJ, Bonard J-M, BGA D. Elastic modulus of ordered and disordered multiwalled carbon nanotubes. Advanced Materials. 1999;**11**(2):161-165. http://ipn2.epfl.ch/CHBU/papers/ourpapers/Salevtat_AdvMater99.pdf

[9] Treacy MMJ, Ebbesen TW, Gibson JM. Exceptionally high Young's Modulus observed for individual carbon nanotube. Nature. 1996;**381**(6584):687-680

[10] Poncharal P, Wang ZL, Ugarte D, De Heer WA. Electrostatic deflections and electromechanical resonances of carbon nanotubes. Science. 1999;**283**(5407):1971-1973. DOI: 10.1126/science.283.5407.1513

[11] Qi HJ, Teo KBK, Lau KKS, et al. Determination of mechanical properties of carbon nanotubes and vertically aligned carbon nanotube forests using nanoindentation. Journal of the Mechanics and Physics of Solids. 2003;**51**(11-12):2213-2237. DOI: 10.1016/j.jmps.2003.09.015

[12] Zhou X, Zhou J, Ou-Yang Z. Strain energy and Young's modulus of single-wall carbon nanotubes calculated from electronic energy-band theory. Physical Review B. 2000;**62**(20):13692-13696. DOI: 10.1103/PhysRevB.62.13692

[13] Pantano A, Parks DM, Boyce MC. Mechanics of deformation of single- and multi-wall carbon nanotubes. Journal of the Mechanics and Physics of Solids. 2004;**52**(4):789-821. DOI: 10.1016/j.jmps.2003.08.004

[14] Chen X, Liu L, Cao G. Mechanisms of nanoindentation on multiwalled carbon nanotube and nanotube cluster. Journal of Nanomaterials. 2008;**2008**:1. DOI: 10.1155/2008/271763

[15] Chiodarelli N, Masahito S, Kashiwagi Y, et al. Measuring the electrical resistivity and contact resistance of vertical carbon nanotube bundles for application as interconnects. Nanotechnology. 2011;**22**(8). DOI: 10.1088/0957-4484/22/8/085302

[16] Deng J, Sun P, Ping Z, Cheng G. Adhesive enhancement improved field emission characteristics of carbon nanotube arrays on energetic ion pre-bombarded Si substrates. 2011;**483**:589-594. DOI: 10.4028/www.scientific.net/KEM.483.589

[17] Xia Z, Liang J. Multiscale Modeling of carbon nanotube adhesion for dry adhesives. Materials Research Society Symposium Proceedings. 2007;**975**:1-8. DOI: 10.1557/PROC-975-0975-DD10-09

[18] Lahiri I, Lahiri D, Jin S, Agarwal A, Choi W. Carbon nanotubes: How strong is their bond with the substrate? ACS Nano. 2011;**5**(2):780-787. DOI: 10.1021/nn102900z

[19] Ageev OA, Bykov AV, Kolomiitsev AS, et al. Study of modification methods of probes for critical-dimension atomic-force microscopy by the deposition of carbon nanotubes. Semiconductors. 2015;**49**(13):1743-1748. DOI: 10.1134/S1063782615130023

[20] Ageev OA, Blinov YF, Il'in OI, et al. Study of the resistive switching of vertically aligned carbon nanotubes by scanning tunneling microscopy. Physics of the Solid State. 2015;**4**: 807-813. DOI: 10.1134/S1063783415040034

[21] Ageev OA, Konoplev BG, Rubashkina MV, Rukomoikin AV, Smirnov VA, Solodovnik MS. Studying the effect of geometric parameters of oriented GaAs nanowhiskers on Young's modulus using atomic force microscopy. Nanotechnologies Russ. 2013;**8**(1-2): 23-28. DOI: 10.1134/S1995078013010023

[22] Ageev OA, Il'in OI, Kolomiitsev AS, et al. Development of a technique for determining Young's modulus of vertically aligned carbon nanotubes using the nanoindentation method. Nanotechnologies Russ. 2012;**7**(1-2):47-53. DOI: 10.1134/S1995078012010028

[23] Ageev OA, Il'in OI, Rubashkina MV, Smirnov VA, Fedotov AA, Tsukanova OG. Determination of the electrical resistivity of vertically aligned carbon nanotubes by scanning probe microscopy. Technical Physics. 2015;**60**(7):1044-1050. DOI: 10.1134/S1063784215070026

[24] Umemura K, Komatsu J, Uchihashi T, et al. Atomic force microscopy of RecA–DNA complexes using a carbon nanotube tip. Biochemical and Biophysical Research Communications. 2001;**281**:390-395. DOI: 10.1006/bbrc.2001.4333

[25] Ageev OA, Balakirev SV, Bykov AV, et al. Development of new metamaterials for advanced element base of micro- and nanoelectronics, and microsystem devices. In: Parinov IA, Chang S-H, Topolov VY, editors. Advanced Materials – Manufacturing, Physics, Mechanics and Applications. Switzerland: Springer International Publishing; 2016. pp. 563-580

[26] Il'ina MV, Il'in OI, Rudyk NN, Konshin AA, Ageev OA. The memristive behavior of non-uniform strained carbon nanotubes. Nanosyst physics. Chem Math. 2018;**9**(1):76-78. DOI: 10.17586/2220-8054-2018-9-1-76-78

[27] Ageev OA, Blinov YF, Il'in OI, et al. Study of the resistive switching of vertically aligned carbon nanotubes by scanning tunneling microscopy. Physics of the Solid State. 2015;**57**(4):825-831. DOI: 10.1134/S1063783415040034

[28] Ageev OA, Ilin OI, Kolomiytsev AS, Rubashkina MV, Smirnov VA, Fedotov AA. Investigation of effect of geometrical parameters of vertically aligned carbon nanotubes on their mechanical properties. Advances in Materials Research. 2014;**894**:355-359. DOI: 10.4028/www.scientific.net/AMR.894.355

[29] Ageev OA, Blinov YF, Il'ina MV, Konoplev BG, Smirnov VA. Resistive switching of vertically aligned carbon nanotubes for advanced Nanoelectronic devices. In: Tiwari A, Mishra YK, Kobayashi H, Turner APF, editors. Intelligent Nanomaterials. 2nd ed. NJ: John Wiley & Sons; 2017. pp. 361-394. DOI: 10.1002/9781119242628.ch11

[30] Il'in OI, Il'ina MV, Rudyk NN, Fedotov AA, Ageev OA. The growth temperature effect on vertically aligned carbon nanotubes parameters. Nanosyst Physics, Chem Math. 2018;**9**(1):92-94. DOI: 10.17586/2220-8054-2018-9-1-92-94

[31] Klimin VS, Il'Ina MV, Il'In OI, Rudyk NN, Ageev OA. Research of influence of the underlayer material on the growth rate of carbon nanotube arrays for manufacturing non-volatile memory elements with high speed. Journal of Physics Conference Series. 2017;**917**(9). DOI: 10.1088/1742-6596/917/9/092023

[32] Bunch JS, Rhodin TN, McEuen PL. Noncontact-AFM imaging of molecular surfaces using single-wall carbon nanotube technology. Nanotechnology. 2004;**15**(2). DOI: 10.1088/0957-4484/15/2/016

[33] Ichimura K, Osawa M, Nomura K, Kataura H. Tunneling spectroscopy on carbon nanotubes using STM. Physics B. 2002;**323**:230-232

[34] Meunier V, Lambin P. Scanning tunneling microscopy and spectroscopy of topological defects in carbon nanotubes. Carbon. 2000;**38**(11):1729-1733. DOI: 10.1016/S0008-6223(99)00296-1

[35] Bobrinetskii II, Nevolin VK, Stroganov AA, Chaplygin YA. Controlling electrical transport through bundles of single-wall carbon nanotubes. Russian MicroElectronics. 2004; **33**(5):292-297. DOI: 10.1023/B:RUMI.0000043045.98569.10

[36] Ageev OA, Blinov YF, Il'in OI, et al. Memristor effect on bundles of vertically aligned carbon nanotubes tested by scanning tunnel microscopy. Technical Physics. 2013;**58**(12): 1831-1836. DOI: 10.1134/S1063784213120025

[37] Di Bartolomeo A, Scarfato A, Giubileo F, et al. A local field emission study of partially aligned carbon-nanotubes by atomic force microscope probe. Carbon. 2007;**45**(15):2957-2971. DOI: 10.1016/j.carbon.2007.09.049

[38] Won H, Willis RF. A STM point-probe method for measuring sheet resistance of ultra-thin metallic films on semiconducting silicon. Surface Science. 2010;**604**(5-6):490-494. DOI: 10.1016/j.susc.2009.11.028

[39] Ilina MV, Blinov YF, Ilin OI, Klimin VS, Ageev OA. Resistive switching of vertically aligned carbon nanotube by a compressive strain. Proceedings of SPIE. 2016;**10224**: 102240U. DOI: 10.1117/12.2266762

[40] Ageev OA, Blinov YF, Ilina MV, Ilin OI, Smirnov VA. Modeling and experimental study of resistive switching in vertically aligned carbon nanotubes. Journal of Physics Conference Series. 2016;**741**:12168. DOI: 10.1088/1742-6596/741/1/012168

[41] Fathi D, Forouzandeh B. Interconnect challenges and carbon nanotube as interconnect in Nano VLSI circuits. In: Marulanda JM, editor. Carbon Nanotubes. Rijeka: InTech; 2010. DOI: 10.5772/39430

[42] Ageev OA, Blinov YF, Il'ina MV, Il'in OI, Smirnov VA, Tsukanova OG. Study of adhesion of vertically aligned carbon nanotubes to a substrate by atomic-force microscopy. Physics of the Solid State. 2016;**58**(2):309-314. DOI: 10.1134/S1063783416020037

[43] Kundalwal SI, Meguid SA, Weng GJ. Strain gradient polarization in graphene. Carbon. 2017;**117**:462-472. DOI: 10.1016/j.carbon.2017.03.013

[44] Ilina MV, Blinov YF, Ilin OI, Rudyk NN, Ageev OA. Piezoelectric effect in non-uniform strained carbon nanotubes. IOP Conf Ser Mater Sci Eng. 2017;**256**:12024. DOI: 10.1088/1757-899X/256/1/012024

[45] Wang X, Tian H, Xie W, et al. Observation of a giant two-dimensional band-piezoelectric effect on biaxial-strained graphene. NPG Asia Materials. 2015;**7**(1):e154. DOI: 10.1038/am.2014.124

[46] Il'ina MV, Il'in OI, Blinov YF, Konshin AA, Konoplev BG, Ageev OA. Piezoelectric response of multi-walled carbon nanotubes. Materials. 2018;**11**:638. DOI: 10.3390/ma11040638

Chapter 5

Characterization of Multiblock (Segmented) Copolyurethane-Imides and Nanocomposites Based Thereof Using AFM, Nanotribology, and Nanoindentation Methods

Tatiana Evgenievna Sukhanova,
Tatyana A. Kuznetsova, Vasilina A. Lapitskaya,
Tatiana I. Zubar, Sergei A. Chizhik,
Milana E. Vylegzhanina, Aleksandr A. Kutin,
Andrey L. Didenko and Valentin M. Svetlichnyi

Additional information is available at the end of the chapter

http://dx.doi.org/10.5772/intechopen.78625

Abstract

This chapter reviews our results on the morphology, tribological, and local mechanical property investigations of new copoly(urethane-imide)s (coPUIs) and nanocomposites based thereof using atomic force microscopy (AFM) and nanoindentation (NI) methods. AFM in the contact mode of lateral forces revealed the presence of different contrast phases on the surface of synthesized films which depends on the chemical structure of monomers used. Single-walled carbon nanotubes (SWCNTs), carbon nanofibers, graphene, tungsten disulfide and tungsten diselenide were introduced into coPUI matrices. Dependencies of microhardness and modulus of elasticity on the depth of indentation have been obtained. It was found that for each synthesized coPUI, there is only one type of carbon nanomaterials that exerts the greatest influence on their characteristics. The improvement of mechanical properties is found to mainly depend on the nature of the polymer matrix and filler. Our results showed that effective methods for improving of tribological characteristics can be either modification by SWCNTs (up to 1 wt.%) or heating at 30°C. Synthesized coPUI films and nanocomposites are very promising materials and can be used as thermoplastic elastomers for tribological applications, and their physical-mechanical properties can be controlled both by temperature and by mechanical action.

Keywords: atomic force microscopy, nanoindentation, nanotribology, copoly(urethane-imide)s, polymer nanocomposites, morphology, local mechanical properties

1. Introduction

For the creation of new-generation polymer materials based on multicomponent systems, it is necessary to study the morphology, mechanical behavior, and various properties of these systems in the submicron and nanoscale range. It is known that surface layers of polymer objects exhibit significant differences in properties as compared with bulk material and play a decisive role in various applications especially if they are used as films and coatings [1].

In recent years, various probe methods have been successfully used for these purposes, among which the most popular are atomic force microscopy (AFM) and nanoindentation (NI) [2, 3]. Atomic force microscopy is an indispensable technique to study the various surface phenomena [4–6]. AFM allows not only to evaluate the roughness of the surface but also recognize the different localized phases and reveal the distribution of friction forces over the surface and evaluate the tribological characteristics of polymer materials in the micro- and nanoscale [6–9].

The aim of this work was to study the morphological, local mechanical, and tribological characteristics of a series of novel multiblock (segmented) copolymers, containing rigid imide blocks and flexible blocks of polyurethanes—copoly(urethane-imide)s (coPUI)s and their nanocomposites—containing carbon nanofillers of different morphologies: single-walled carbon nanotubes (SWCNTs), vapor-grown carbon nanofibers (VGCF), and graphene as well as nanoparticles of transition metal chalcogenides—tungsten disulfide (WS_2) and tungsten diselenide (WSe_2). The synthesized materials were characterized down to the nanoscale level using AFM in contact and tapping modes, including the friction coefficient measurement [10] at the temperature range from 20 up to 120°C and multi-pass friction force determination till 400 scanning fields, and NI method [11] for the local mechanical property evaluation as the dependence of the characteristics on the depth of an indenter penetration.

Multiblock (segmented) copolymers are the block copolymers containing a rigid block of monomer A and a flexible-chain polymer block (B)k ([A–(B)k]n). A characteristic feature of such copolymers is the microphase separation of blocks A and (B)k, since blocks A and (B)k are thermodynamically incompatible [12]. As a result, nano- and micro-regions are formed in the volume of the copolymer, in which segments (blocks) of the same chemical structure are concentrated. Thanks to the microphase separation of the rigid blocks (A) and flexible blocks of aliphatic ethers ((B)k), these copolymers acquire the properties of elastomers: their glass transition temperature (T_g) is in the negative temperature range on the Celsius scale, and the temperature dependencies of the elastic modulus at higher T_g are distinguished by a wide range of rubberlike elasticities, while the relative tensile elongation at room temperature is hundreds of percent [13].

Changing the chemical structure and the ratio of rigid and flexible blocks in coPUIs, one can change the structure, morphology, and mechanical and thermal properties of these polymers [14–17]. Modification of multiblock coPUI with the introduction of carbon and chalcogenide nanosized additives aims to expand their temperature range without losing its elasticity. Finally, the new generation of thermoplastic elastomers for working in the extremal conditions based on multiblock coPUIs intended to be prepared which exhibit unique rubberlike elasticity in a wide range of temperatures and a higher heat resistance than thermoplastic polyurethanes [6, 9, 18–20]. Such materials are very promising for application in airspace technique, shipbuilding and car industry, microelectronics, and membrane technology [21, 22].

2. Experimental details

2.1. Materials

Chemical approaches to the synthesis of multiblock (segmented) copolyester-imides and copolyurethane-imides (coPUIs) are proposed in general form in previous works [23–25]. In these studies, macromonomers with terminal anhydride groups are used which are formed by terminating with tetracarboxylic acid dianhydrides of polyesters having terminal hydroxyl groups (in the case of copolyester-imides) or terminal isocyanate groups (in the case of copolyurethane-imides), respectively. Further, the expansion of the macromonomer chains is carried out using aromatic diamines which interact with their terminal anhydride groups. The polyacylation of diamines with macromonomer dianhydrides is carried out according to schemes known in the chemistry of polyimides.

For the synthesis of new family of multiblock coPUIs, the 1,3-bis-(3′,4-dicarboxy-phenoxy)benzene dianhydride (R) (T_m = 163–165°C, LLC TekhChemProm, Yaroslavl) and diamine1,4′-bis-(4′-aminophenoxy)diphenyl sulfone (diamine SOD) (T_m = 194–196°C, VWR International) with various ester fragments, as well as an aliphatic copolyester-poly(propylene glycol) (PPG) or poly(1,6-hexanediol/neopentyl glycol-alt-adipic acid) (Alt) (900) (T_m = 33°C, M_n = 900, Aldrich), secondarily terminated with 2,4-tolylene diisocyanate (TDI 2300) (T_m = 19.5–21.5°C, M_n = 2300, Aldrich), were used. The chemical structures of the coPUIs are presented in **Schemes 1** and **2**.

Depending on the chemical structure of flexible aliphatic and rigid aromatic blocks, the synthesis of copolymers was carried out according to different schemes.

Synthesis of multiblock (segmented) coPUI (R-AltTDI-R)SOD (**Scheme 1**) includes four stages:

1. The stage of formation of a macromonomer with terminal toluene isocyanate groups (AltTDI).
2. The stage of formation of a macromonomer with terminal anhydride groups (R-AltTDI-R).
3. The step of polyacylation of diamine (expansion of the macromonomer chain) to form a prepolymer-copolyamide acid (R-AltTDI-R)SOD.
4. The stage of thermal imidization of the prepolymer (copolyamide acid) to form copolyurethane-imide (R-AltTDI-R)SOD.

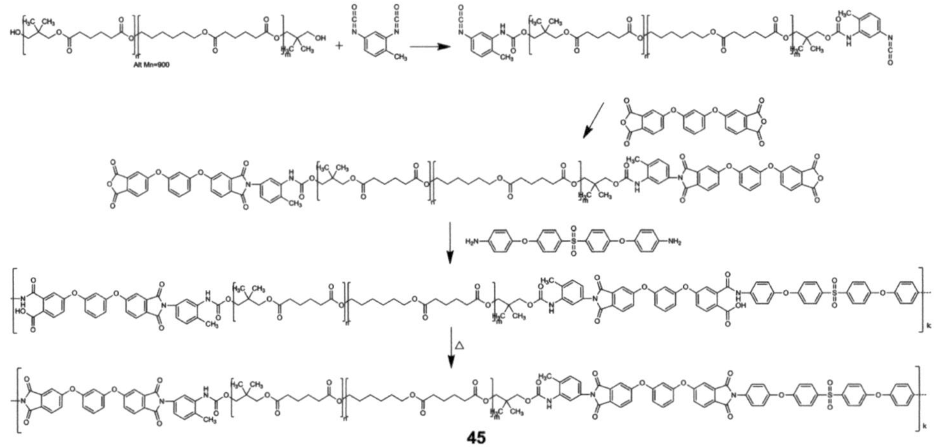

45

Scheme 1. Schematic representation synthesis of coPUI (R-AltTDI-R)SOD.

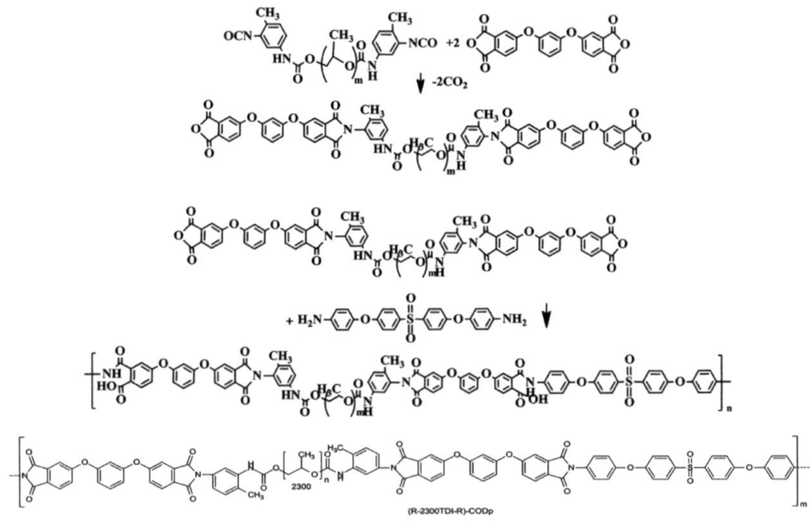

Scheme 2. Schematic representation synthesis of coPUI (R-2300TDI-R)SOD.

Synthesis of multiblock (segmented) coPUI (R-2300TDI-R)SOD includes three stages (**Scheme 2**) as described previously in our papers [6, 19, 23]:

1. The step of forming a macromonomer with terminal anhydride groups (R-2300TDI-R).
2. The step of diamine polyacylation (expansion of the macromonomer chain) to form a prepolymer-copolyamide acid (R-2300TDI-R)SOD.
3. The stage of thermal imidization of the prepolymer to form coPUI (R-2300TDI-R)SOD.

Thermal imidization of the obtained prepolymers in N-methyl-2-pyrrolidone (NMP) solution was carried out in an inert atmosphere (argon) at 165°C for 3.5 h. Dean-Stark packing was used to remove the water that was emitted from the imidization (curing of the prepolymer) in the form of an azeotropic mixture with toluene. The temperature was then raised to 210°C to complete the imidization process.

To obtain nanocomposites, the polymer solutions prepared during the synthesis were used. In the first step, the prepared nanoparticle powder solutions in NMP—graphene, vapor-grown carbon nanofibers (VGCF), single-walled carbon nanotubes (SWCNTs), tungsten disulfide, and tungsten diselenide (WS_2 and WSe_2, correspondingly)—were dispersed in NMP by ultrasound. Then, synthesized polymer solution was added and mixed on a mechanical stirrer for 4 h. The nanoparticles are taken so that, after removal of the solvent, the amount of nanoparticles in the polymer is the required amount of percent by weight. Solutions of the resulting mixture were applied to a glass, polyimide (PI) film, fluoroplastic, and aluminum substrates. The resulting polymer coatings were dried at 80°C for 12 h and, then, heated stepwise at 120, 140, 160, and 180°C for 1 h at each step. The thickness of the heated films ranged from 100 to 300 μm.

It is known that imide-containing film coatings display strong adhesion to various substrates [26]. Some film samples were prepared on aluminum substrates. The coating samples formed on aluminum substrates were immersed in concentrated hydrochloric acid to remove the substrate, and then the resulting films were repeatedly washed in distilled water. The nano-composite films based thereon containing from 1 to 10 wt.% carbon nanofillers of different morphologies—SWCNT, VGCF, graphene, and WS_2 and WSe_2 nanoparticles—were prepared.

2.2. Methods

The surface morphology of the synthesized copolymer films was investigated by the AFM method on the AFM device NT-206 (produced by "Microtestmashiny ODO," Belarus) using standard CSC38 and NSC11-type silicon probes with the radius of tip curvature less than 10 nm and stiffness of cantilever 0.08 and 3 N/m produced by "MikroMasch" (Estonia). The roughness, coefficient of friction, and friction force in nano- and microscale of samples were evaluated. The surface roughness was estimated in scanning areas of 2 × 2, 5 × 5, and 10 × 10 μm.

The friction force and friction coefficient (C_{fr}) were determined using 20 × 20 μm scanning areas with silicon probe NSC11 of V-shaped type with stiffness of cantilever 3 N/m at the moving speed 1.55 μm/s. The tip radius was specially blunted by pre-scanning to curvature 100 nm. Such preparation prevents the probe from further blunting and keeps the contact condition stable [8]. C_{fr} is calculated from a ratio of the force of friction between two bodies and the normal force pressing them together and their measurement as described previously [8, 10]. To measure the friction force, the AFM probe scans the surface in the so-called two-pass method [10]. During the first pass, the probe goes forward along the surface. When the probe returns back in the second pass, it moves according to the first line, which suits to the relief of the surface. During the forward and the back passes, the friction forces act from the surface on a tip, which cause the torsion of the cantilever besides the cantilever's deflection in the normal direction. The angle of this twist depends on the value of friction force. To calculate the effect of frictional forces more precisely, the angle of twist when the probe is moved in the forward and backward directions is divided in half [7, 8].

C_{fr} was determined in several regimes: on one scan field for one line in the forward and backward direction and so on the multiple scans (200–400) in one place. The measurements were carried out both at room temperature and at a temperature of 120°C. The heating was carried out directly during the scanning process using a thermoplatform placed on the AFM table. The thermoplatform is equipped with heaters and a feedback system with sensors, which allows controlling the heating and automatically reducing the voltage, if necessary, to maintain the set temperature.

To determine the local mechanical properties of prepared films based on multiblock coPUI, a 750 Ubi (Hysitron, USA) nanoindentator with a Berkovich indenter with a radius of curvature of R = 100 nm was used. Berkovich indenters are preferred than the Vickers geometry indenters because the latter is prone to undesired "roof" tip imperfections.

Since the physical-mechanical properties of polymers depend not only on the chemical composition and macromolecule structure but also on their orientation in the material and the distribution of the unordered amorphous and partially ordered crystalline phases, they cannot be characterized by a single value of the elasticity modulus (E) or microhardness (H) [11]. In this study, the local mechanical properties were analyzed as a dependence of the characteristic on the depth of penetration into material. For each sample, not less than 100 indentation curves were performed with an increasing load of 50–5000 μN in 50 μN increments, and the dependencies of the E and H on the depth of indentation were constructed from the results obtained.

In order to compensate the size effect of nanoindentation at shallow depths associated with the fact that at small penetration depths the Berkovich tip has the shape of a hemisphere rather than a pyramid, the nanoindentator was pre-calibrated according to a standard sample of fused quartz in a range of loads that were individually selected for each sample by the corresponding depths from 5 to 200 nm. This approach allows achieving a minimum standard deviation in the measurements of E and H.

3. Results and discussion

3.1. Carbon nanoadditive influence on the coPUI film morphology

Typical AFM 3D images of the coPUI matrices and nanocomposite surfaces are presented in **Figures 1** and **2**. It is clearly seen that both sides—free side and bottom side—of the synthesized films significantly differ in the surface morphology. It can be concluded that their morphology and, also, the structure are extremely sensitive to the substrate surface nature as well as to the formation process. Hence, the effect of additive (graphene and SWCNTs) influence on the film structure is also large, which is confirmed by AFM images (**Figures 1** and **2**). In our work, special attention was paid to the definition of roughness parameters: R_a, arithmetic mean; R_q, rms surface roughness.

According to the AFM image, free surface of coPUI (R-2300TDI-R)SOD film (**Figure 1a**) at scanning area of 1.5 × 1.5 μm is smooth (R_a = 1.0 nm, R_q = 1.3 nm) and exhibits grain morphology with a lot of pore nano- and meso-sizes. For scanning area of 2 × 2 μm, this film surface has a roughness R_a = 3.2 nm and R_q = 4.9 nm on the free surface and R_a = 3.1 nm and R_q = 4.0 nm on the surface adjacent to the substrate.

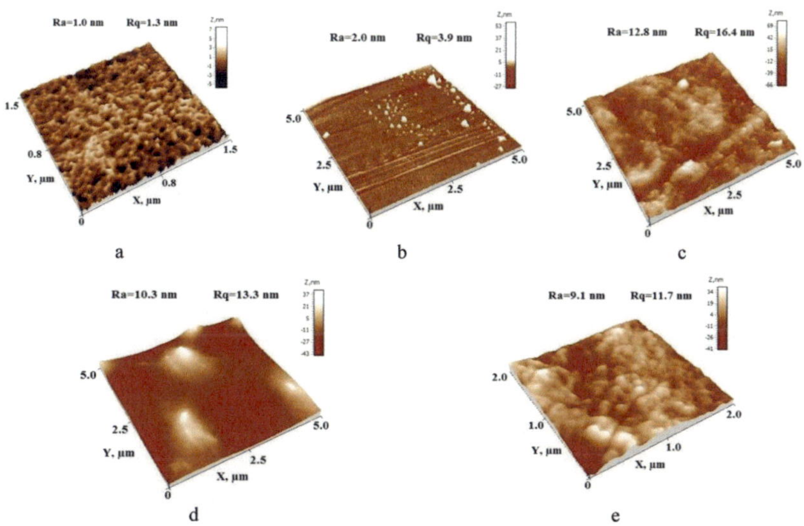

Figure 1. 3D images of (a) the coPUI(R-2300TDI-R)SOD matrix (side 1, scanning area 1.5 × 1.5 μm) and nanocomposites based thereon, (b) with 1 wt.% graphene (side 1, scanning area 5 × 5 μm), (c) with 1 wt.% graphene (side 2, scanning area 5 × 5 μm), (d) with 1 wt.% SWCNTs (side 1, scanning area 5 × 5 μm), and (e) with 1 wt.% SWCNTs (side 2, scanning area 2 × 2 μm). Glass substrate.

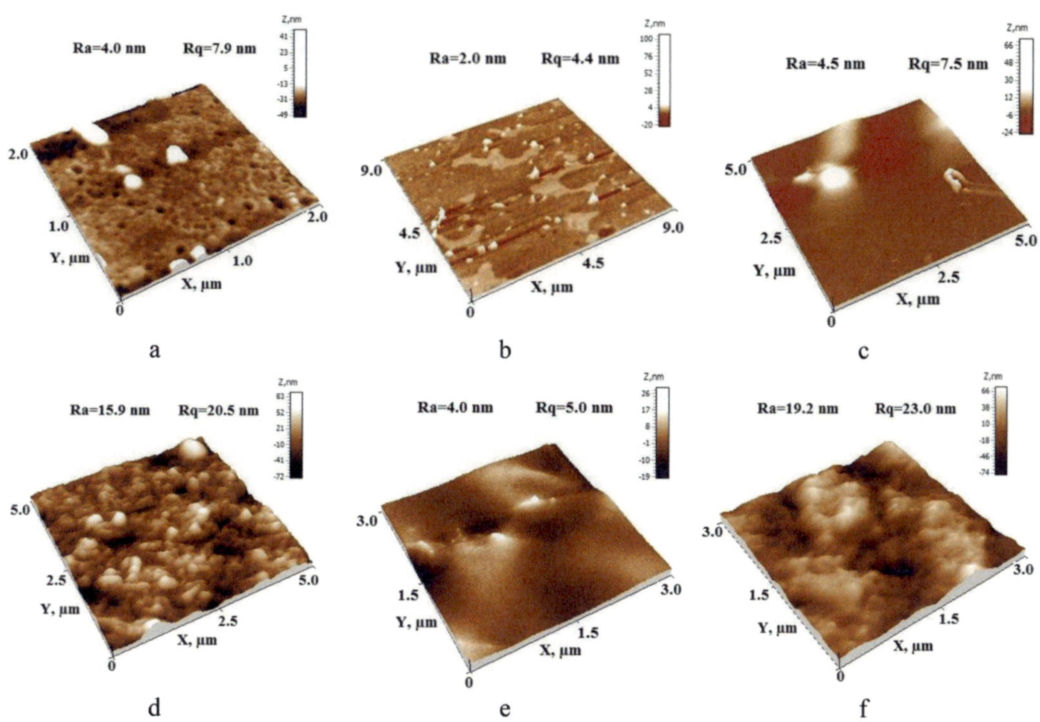

Figure 2. 3D images of the nanocomposites based on (R-AltTDI-R)SOD matrix: (a) matrix (side 1, scanning area 2 × 2 μm), (b) matrix with phase separation (side 1, scanning area 9 × 9 μm), (c) matrix with 1 wt.% VGCF (side 1, scanning area 5 × 5 μm), (d) matrix with 1 wt.% VGCF (side 2, scanning area 5 × 5 μm), (e) matrix with 1 wt.% SWCNTs (side 1, scanning area 3 × 3 μm), and (f) matrix with 1 wt.% SWCNT (side 2, scanning area 3 × 3 μm). Glass substrate.

For nanocomposites based on coPUI (R-2300TDI-R)SOD containing 1 wt.% graphene (**Figure 1b,c**), the great difference in morphology between free surface and surface adjacent to the substrate has been observed. The surface which was in contact with substrate (**Figure 1c**) looks more rough with roughness parameters R_a = 12.8 nm and R_q = 16.4 nm (for scanning area 5 × 5 μm) and R_a = 10.1 nm and R_q = 12.5 nm (for scanning area 2 × 2 μm). On the free surface of nanocomposite with 1 wt.% SWCNTs (**Figure 1d**), the ends of carbon nanotubes can be recognized. Again, the surface adjacent to the substrate is still rough and does not have homogeneous morphology like the free one (**Figure 1e**).

The significant difference in the topography of both coPUI nanocomposite film surfaces is clearly visible when comparing images in **Figure 1b–e**. Also, these surfaces exhibit a significant difference in the roughness values (**Table 1**).

Nanocomposite films of (R-2300TDI-R)SOD +1 wt.% graphene (**Figure 1b,c**) display the different roughness values between the film surfaces. When adding graphene, the free surface relief becomes coarser. It is likely that graphene nanoparticles are concentrated on one of the surfaces.

On the other hand, nanocomposite film of (R-2300TDI-R)SOD +1 wt.% SWCNT (**Figure 1d** and **e**) displays the close roughness values on both surfaces: R_a = 26.1 nm on the free surface and R_a = 26.5 nm on the surface adjacent to the substrate (for scanning area 5 × 5 μm). When the scanning area is reduced to 2 × 2 μm, the roughness values become several times smaller: R_a = 6.3 nm on the free surface and R_a = 11.9 nm on the surface adjacent to the substrate (**Table 1**). It can be concluded that the SWCNT is homogeneously distributed in composite volume.

For coPUI (R-2300TDI-R)SOD with 1 wt.% VGCF, the different roughness values between both surfaces were determined: R_a = 5.1 nm on the free surface and R_a = 23.1 nm on surface adjacent to the substrate (for scanning area 5 × 5 μm). When the scan area is reduced to 2 × 2 μm, the roughness values become R_a = 1.2 nm and R_q = 1.5 nm on the free surface and R_a = 13.8 nm and R_q = 18.0 nm on the surface adjacent to the substrate (**Table 1**).

Comparison of the roughness values for coPUI(R-2300TDI-R)SOD +1 wt.% VGCF films with all films of nanocomposites based on coPUI (R-AltTDI-R)SOD showed that they are about three times different (**Table 1**). Thus, additives significantly influence on the morphology formation of these coPUI films and on their surface properties.

The example of the phase separation on the free surface of coPUI (R-AltTDI-R)SOD film is clearly seen in **Figure 2b**. In the mode of lateral force contrast, light spots surrounded by dark polymer matrix have been observed that indicates a significant difference in the tribological properties of the phases in this sample.

3.2. Substrate nature influence on the coPUI film morphology

Substrate nature influence on the polymer film morphology can be traced comparing the surface morphology of coPUI (R-AltTDI-R) films prepared on three types of substrates: on a glass substrate, on the PI film substrate, and on a fluoroplastic substrate (**Figures 3–5**).

Figure 3 shows the AFM images of free surface (a–c) and surface adjacent to substrate (d–f) cast on a glass. A large number of pores with diameters from 50 to 200 nm are observed on both

Polymer	Samples symbol	Film side	Scanning area (2 × 2 μm²)		Friction coefficients
			Roughness (R_a, nm)	Roughness (R_q, nm)	
(R-2300TDI-R)SOD {50}	50	1	3.2	4.9	0.164
		2	3.1	4.0	0.167
	50 + 1 wt.% VGCF	1	1.2	1.5	0.074
		2	13.8	18.0	0.061
	50 + 1 wt.% SWCNT	1	6.3	8.0	0.183
		2	11.9	14.9	0.185
	50 + 1 wt.% graphene	1	5.8	7.5	0.382
		2	10.1	12.5	0.170
	50 + 3 wt.% WS_2	1	13.5	16.9	0.262
		2	16.9	23.1	0.157
	50 + 1 wt.% WSe_2	1	9.3	14.2	0.158
		2	12.2	15.8	0.068
(R-AltTDI-R)SOD {45}	45	1	5.2	9.6	0.055
		2	6.2	7.9	0.065
	45 + 10 wt.% WS_2	1	3.4	4.9	0.179
		2	13.9	16.5	0.101
	45 + 10 wt.% WSe_2	1	16.5	21.9	0.107
		2	17.5	21.3	0.133
	45 + 1 wt.% VGCF	1	10.8	13.2	0.066
		2	14.1	18.1	0.071
	45+ 1 wt.% SWCNT	1	6.8	11.4	0.065
		2	25.4	31.1	0.063
	45 + 10 wt.% graphene	1	52.1	64.6	—
		2	14.4	18.2	0.092

Table 1. Comparison of the roughness and friction coefficient measurements using AFM in the contact mode for films and nanocomposites casted on a glass substrate.

surfaces, the pore depth at the free surface according to the profile (b) reaches ~2–8 nm, and at the surface to a glass substrate (e), it equals to 2–4 nm. The surface roughness differences are twofold (R_a = 0.9 nm and R_q = 1.1 nm for the surface adjacent to the substrate and R_a = 1.9 nm and R_q = 2.4 nm for the free surface). Therefore, both sides of the film look very similar.

Unlike a glass substrate, PI film substrate significantly changes surface topography (**Figure 4**). Thus, a lot of fine pores are not observed on the free surface. The surface adjacent to the substrate (d–f) is sufficiently smooth (R_a = 0.2 nm and R_q = 0.3 nm); a nanodomain morphology and a rather large number of nanopores with a depth not exceeding 1 nm are observed.

In **Figure 5**, AFM images of the film cast on the fluoroplastic substrate are given. The free surface (a–c) becomes more loose and nonuniform, and the roughness value (R_a = 5.8 nm and

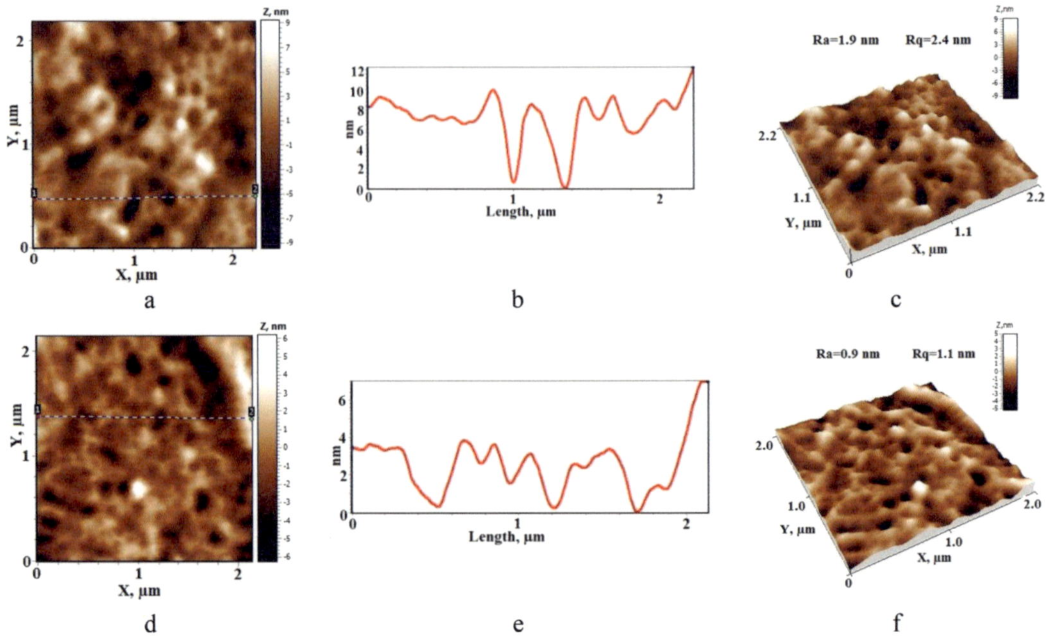

Figure 3. AFM images of coPUI (R-AltTDI-R)SOD film cast on a glass substrate: free surface (a–c) and surface adjacent to the substrate (d–f); (a,d) height images, (b,e) profile, and (c,f) 3D image.

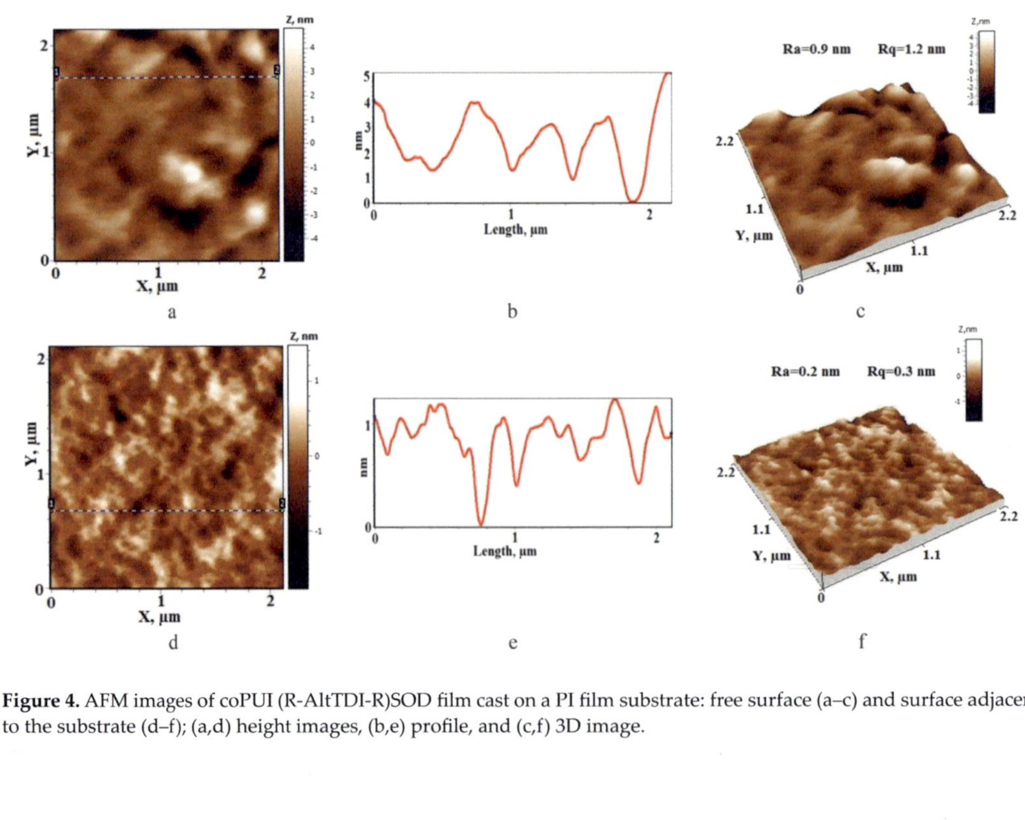

Figure 4. AFM images of coPUI (R-AltTDI-R)SOD film cast on a PI film substrate: free surface (a–c) and surface adjacent to the substrate (d–f); (a,d) height images, (b,e) profile, and (c,f) 3D image.

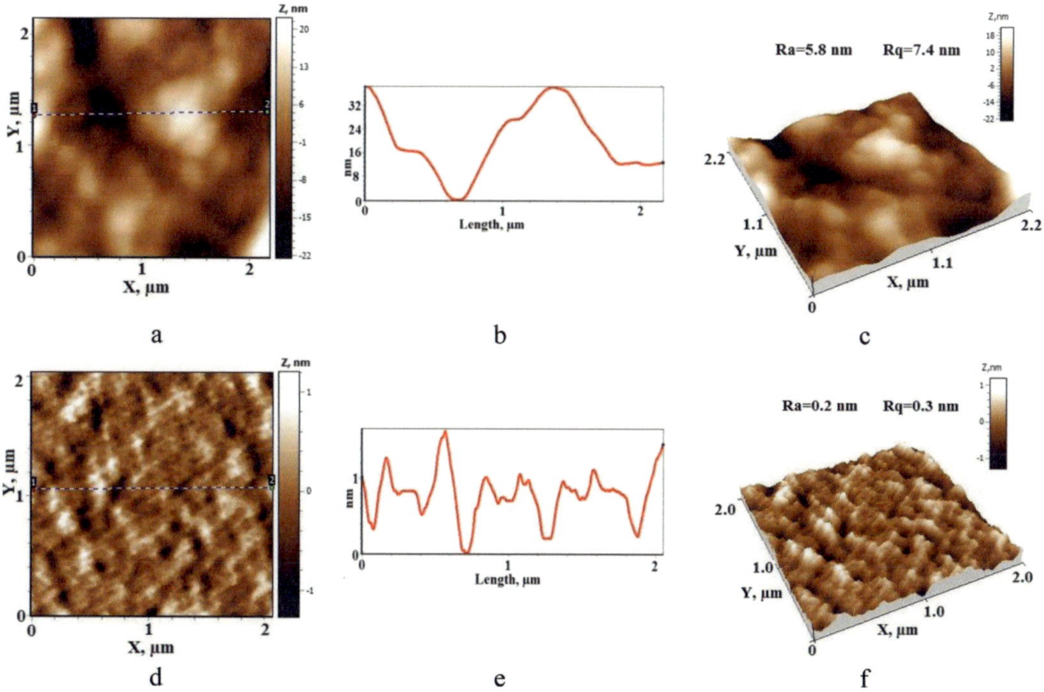

Figure 5. AFM images of coPUI (R-AltTDI-R)SOD film cast on a fluoroplastic substrate: free surface (a–c) and surface adjacent to the substrate (d–f); (a,d) height images, (b,e) profile, and (c,f) 3D image.

R_q = 7.4 nm) is significantly increased. As in the previous case, the surface adjacent to the substrate (d–f) has a nanodomain morphology and a large number of nanopores. Thus, it can be concluded that the nature of the substrate material on which the films are cast has a significant effect on the morphology of the sample.

3.3. Tribological characteristics of the coPUI films and nanocomposites

Friction coefficients (C_{fr}) of coPUIs based on (R-2300TDI-R)SOD and (R-AltTDI-R)SOD matrices modified by carbon nanomaterials determined by AFM method on one field are presented in **Figure 6**. C_{fr} of matrices without additives differ threefold: 0.164–0.167 for (R-2300TDI-R)SOD and 0.056–0.065 for (R-AltTDI-R)SOD.

The addition of VGCF reduces C_{fr} of (R-2300TDI-R)SOD matrix more than twice to 0.061–0.074 and leaves C_{fr} of (R-AltTDI-R)SOD matrix at the low level about 0.066–0.071. The addition of SWCNT practically leaves C_{fr} of both matrices on the level of C_{fr} of pure matrices.

On the other hand, the addition of graphene increases C_{fr} of both matrices especially of (R-2300TDI-R)SOD. Thus, only the addition of VGCF to (R-2300TDI-R)SOD matrix improves the tribological properties of the surface. The excess of the quantity necessary for the modification, as in the case of coPUI with 10 wt.% graphene, can substantially worsen the surface properties.

Friction coefficients of coPUIs based on (R-2300TDI-R)SOD and (R-AltTDI-R)SOD matrices modified by WS_2 and WSe_2 nanoparticles determined by AFM method on one field are

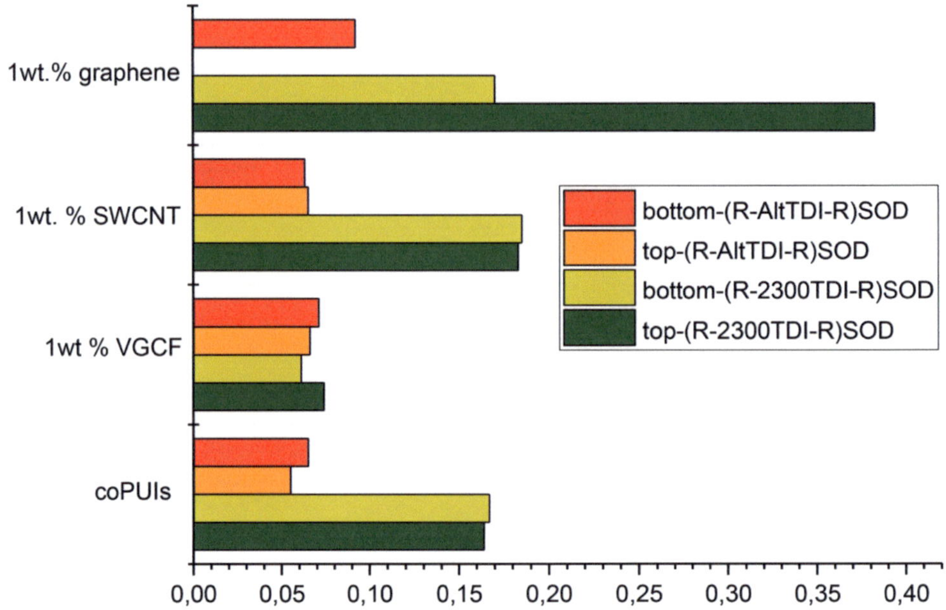

Figure 6. Friction coefficients of coPUIs based on (R-2300TDI-R)SOD (50) and (R-AltTDI-R)SOD (45) matrices and composites with carbon nanofillers as determined by AFM method in the contact mode on one field: matrices 50 and 45 without additives, matrices with 1 wt.% VGCF, matrices with 1 wt.% SWCNTs, and matrices with 1 wt.% graphene.

presented on **Figure 7**. In general, WS_2 and WSe_2 nanoparticles degrade C_{fr} of the (R-2300TDI-R) SOD and (R-AltTDI-R)SOD matrices due to a significant increase in roughness. WSe_2 decreases twice the C_{fr} of (R-AltTDI-R)SOD matrix surface casted on a glass substrate.

The results of C_{fr} changing during multi-pass tribological AFM testing of coPUI (R-AltTDI-R) SOD matrix as a dependence of "C_{fr}—number of scans" are presented in **Figure 8** (black solid line). This measurement was performed at 20°C and normal load about 175 nN. Friction forces were determined in a range of 10–160 nN. Each point on the C_{fr} graph (black solid line) and the roughness graph (blue solid line) was obtained from a separate AFM sequential scanning field (**Figure 8**).

At the initial stage of C_{fr} graph up to 50 cycles, the growth of C_{fr} to a value 0.9 was determined (black solid line), and the roughness values simultaneously decreased (blue solid line). When the probe moves along the surface, the roughness is smoothed out (blue solid line). The low initial value of C_{fr} 0.1 can be explained by the fact of developed roughness. In this cause the contact of the probe with the surface occurs along the tops of the irregularities. In this case, the contact area of the probe and the surface are composed of a plurality of point contacts, and the frictional forces are lower than in case when the smoothed sample contacts with entire surface of the probe. The energy should be spent on the process of smoothing, and it needs to apply a greater lateral force for further movement. Therefore, C_{fr} grows on this side. In the process of friction, the irregularities are smoothed out, and after reaching a minimum of roughness, the C_{fr} decreases over the next 100 cycles to a stable value of 0.05 (black solid line). And then, the mechanism of elastic non-wear friction works.

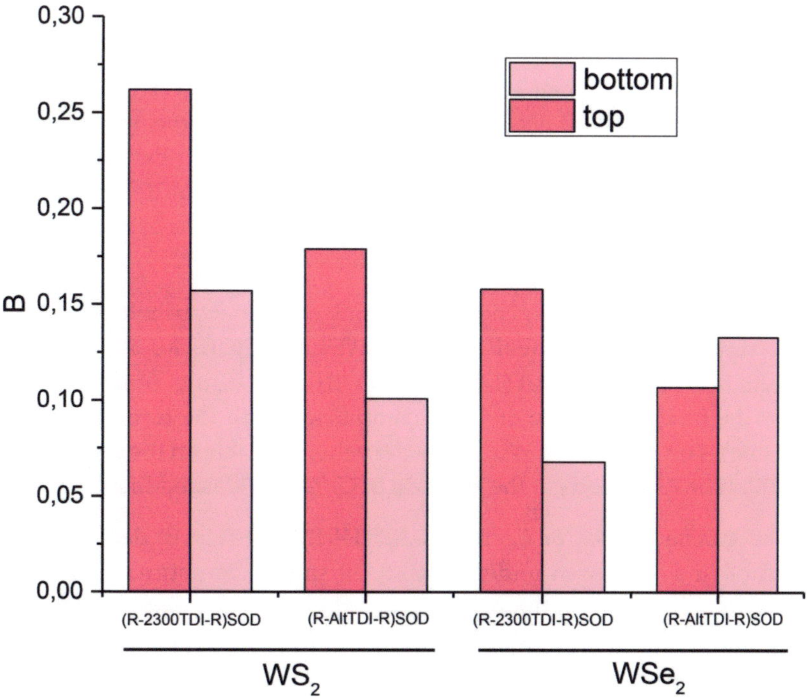

Figure 7. Friction coefficients of coPUIs based on (R-2300TDI-R)SOD (50) and (R-AltTDI-R)SOD (45) matrices modified by WS_2 and WSe_2 nanoparticles, determined by AFM method in the contact mode on one field.

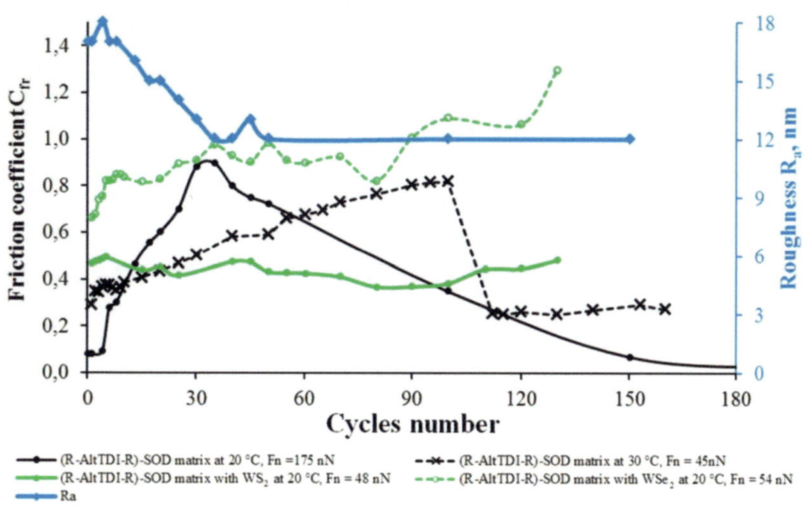

Figure 8. The results of coefficient of friction (C_{fr}) and roughness (R_a) changing during multi-pass tribological AFM testing of (R-AltTDI-R)SOD matrix: C_{fr} of pure matrix at 20°C and F normal = 175 nN (black solid line), C_{fr} of pure matrix at 30°C and F normal = 45 nN (black-dashed line), C_{fr} of matrix with WS_2 at 20°C and F normal = 48 nN (green solid line), C_{fr} of matrix with WSe_2 at 20°C and F normal = 54 nN (green-dashed line), and roughness (R_a) of pure matrix at 20°C and F normal = 175 nN (blue solid line).

When the same multi-pass tests are repeated at the room temperature about 30°C, the probe-sample contact conditions change, which raises the initial value of C_{fr} to 0.3 instead of 0.1 at 20°C (**Figure 8**, black-dashed line). Then, normal load was about 116 nN, and friction forces were determined in range 30–92 nN. The further stages of the curve also change somewhat. C_{fr} is stable at the beginning during the first 10 cycles, and then it increases smoothly to 0.8 (black solid line). After reaching this value at 110 cycles as well as at 20°C, the curve decreases to a stable level about 0.25 (black solid line). The testing at 30°C places peculiarities on the C_{fr} graph: the growth of values is smoother, and the decrease is faster.

The results of coefficient of friction changing during multi-pass tribological AFM testing at 30°C of (R-AltTDI-R)SOD matrix with WS_2 and WSe_2 nanoparticles are shown in **Figure 8** (green lines). As for the pure (R-AltTDI-R)SOD matrix, the results of testing at higher temperature show the increased value of C_{fr} in comparison with the testing at 30°C (**Figure 8**, black-dashed line). The addition of WS_2 keeps the value C_{fr} stable on the level 0.45 (green solid line). The addition of WSe_2 leads to the increase of C_{fr} (green-dashed line).

Comparing the similar values of C_{fr} for (R-AltTDI-R)SOD matrix determined in regimes "multi-pass scan" and "one scan under heating," it should be noticed that the influence of heating in a range of 75–85°C on the surface properties is similar to the influence of 200–400 scans of multi-pass testing at 20°C [8]. This fact allows to make a conclusion that there is heating in this micro-contact equal to 75°C.

These results show that synthesized modified coPUI films are very promising for tribological applications, and their properties can be controlled both by temperature and by mechanical action.

The influence of heating on C_{fr} for (R-AltTDI-R)SOD with WSe_2 addition has been shown in our previous work [8]. It was found that C_{fr} keeps stable at the level of 0.30–0.35 till 45°C. Then, it dramatically decreases up to 0.05 at 60°C and keeps stable till 70°C. Again, at 75°C it decreases till 0.01.

3.4. Nanoindentation of coPUI films

On the dependence of E on the depth of indentation for coPUI (R-2300TDI-R)SOD, two curves are characteristic for the material. It can be concluded that there is a less elastic phase on the surface layer and a more elastic phase in the depth—in bulk material (**Figure 9**, blue curves). In this case, the microhardness for coPUI (R-2300TDI-R)SOD has a unimodal distribution in values from 2 to 4 MPa, and one curve (**Figure 9**, red curves) is present on the plots of the microhardness versus the depth of indentation. Different markers denote the values obtained from different sides of the films. For microhardness they are very close, and for the elastic modulus completely coincide.

On the other hand, coPUI (R-AltTDI-R)SOD is characterized by unimodal distributions for both E and H (**Figure 10**). The transition from the surface to the depth of the sample decreases both for E from 6 to 3 GPa and for H from 400 to 200 MPa.

The performed studies of the samples of synthesized coPUI showed that the values of H and E measured on both sides of the films are practically the same, and when the dependencies obtained from different sides of the films coincide in one plot, for each coPUI they almost

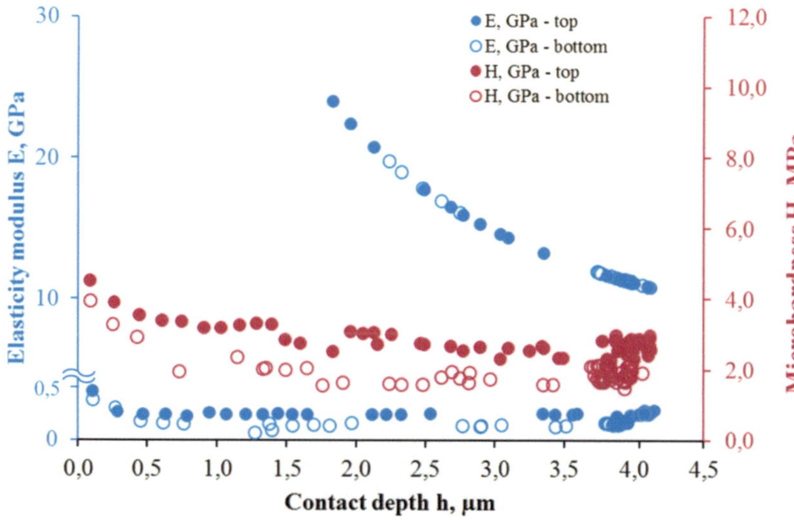

Figure 9. Dependencies of elasticity modulus (blue curves) and microhardness (red curves) versus the depth of indentation for the coPUI (R-2300TDI-R)SOD without nanofillers.

completely coincide, forming exponential curves. The values of E and H for film cast on glass and fluoroplastic substrates are identical, despite the different microstructure of the surface, so hereinafter we consider the values obtained for films prepared on a glass substrate.

With the addition of 1 wt. % SWCNTs, the initial values of E and H for coPUI (R-2300TDI-R) SOD practically do not change (**Figure 11**). For the "amorphous" phase on top of the film E is 25 MPa and from below 20 MPa. At the same time, H is on the top of 3 MPa and from the bottom 2 MPa. For the "partially ordered" phase, E varies from 11 to 14 GPa from the top, from 10 to 22 GPa from the bottom, and from the top and H from below from 1.5 to

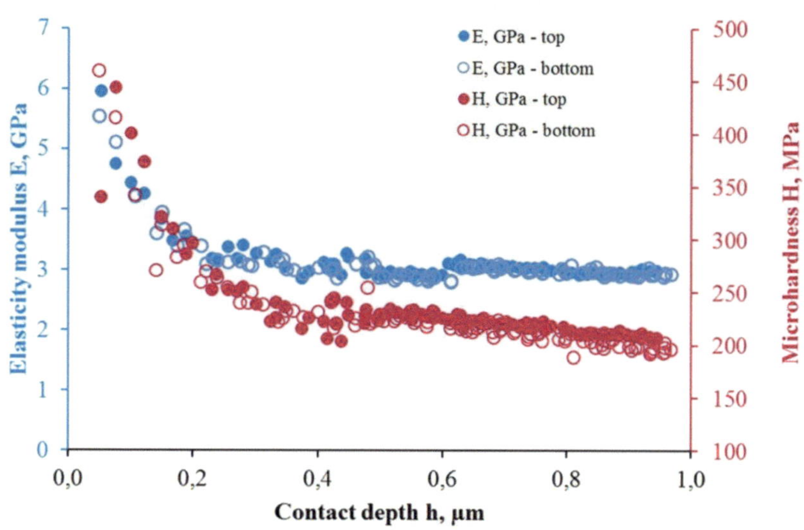

Figure 10. Dependencies of elasticity modulus (blue curves) and microhardness (red curves) versus the depth of indentation for the coPUI (R-AltTDI-R)SOD without nanofillers.

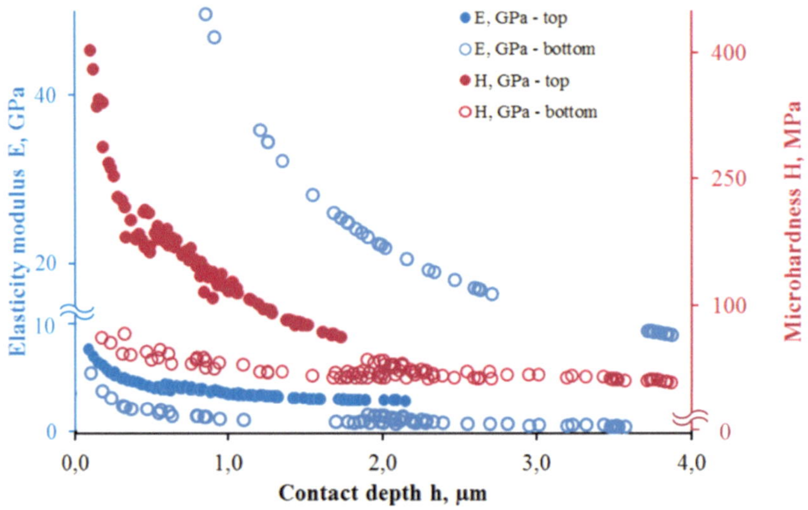

Figure 11. Dependencies of elasticity modulus (blue curves) and microhardness (red curves) versus the depth of indentation for the coPUI (R-2300TDI-R)SOD with 1 wt.% SWCNTs.

2.5 MPa. The microstructure of the surface remains unstructured, which is characteristic for polymers in the amorphous state.

The amorphous phase on the upper surface of the film practically does not change. The depth of the "partially ordered" phase changes: it moves closer to the free surface (**Figure 11**). From below, the depth of the location of the "partially ordered" phase remained unchanged. For an amorphous phase from below, the values of E and H increased by a factor of 2: E was 20 MPa and became 40 MPa, and H was 2 MPa and became 4 MPa.

The greatest changes in the properties of coPUI (R-2300TDI-R)SOD were detected with the addition of 1 wt.% SWCNTs. The "partially ordered" phase (or moved to a depth greater than 4 µm) completely disappeared from the top, and from the bottom, it was preserved and moved closer to the surface than in the coPUI without SWCNTs (**Figure 12a**). The E is 10–20 GPa. For an amorphous phase, the E film increased 10 times from the top—it was 35 MPa and became 350 MPa. The H of the amorphous phase increased 23-fold: from 3 to 70 MPa. From the bottom of the film, the values of H and E increased by 5 and 10 times: E was 20 MPa and became 100 MPa, and H was 2 MPa and became 20 MPa. The microstructure of the surface is structured, with noticeable linear formations in perpendicular directions in the plane (**Figure 12d** and **e**).

The introduction of all types of nanofillers in an amount of 1 wt.% practically did not change the mechanical properties of the coPUI (R-AltTDI-R)SOD. In our experiments only the addition of graphene in an amount about 10 wt.% lowered the properties of this for coPUI (R-AltTDI-R)SOD by a factor of 1.5. In general, the material with 10 wt.% graphene is unstable due to the accumulation of excess graphene in the form of submicron-sized inclusions. The surface microstructure of the composite became nanostructured.

As a result, it was established that the coPUI (R-2300TDI-R)SOD and coPUI (R-AltTDI-R)SOD are radically different in the distribution of the elasticity modulus E: coPUI (R-2300TDI-R)SOD has a bimodal distribution (**Figure 13a**), and for coPUI (R-AltTDI-R)SOD, it is unimodal (**Figure 13b**).

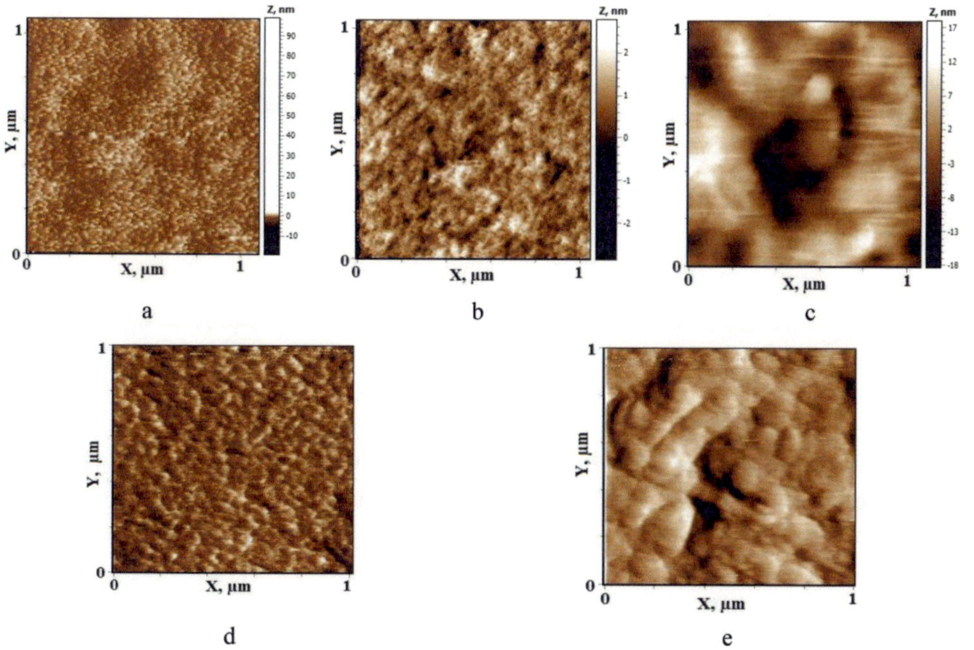

Figure 12. AFM images of the surface microstructure of the coPUI (R-2300TDI-R)SOD: (a) the initial matrix (free side), (b) matrix with 1 wt.% graphene (free side), (c) the side from the substrate, (d) matrix with 1 wt.% SWCNTs (free side), and (e) side from the substrate; scan area 1 × 1 μm.

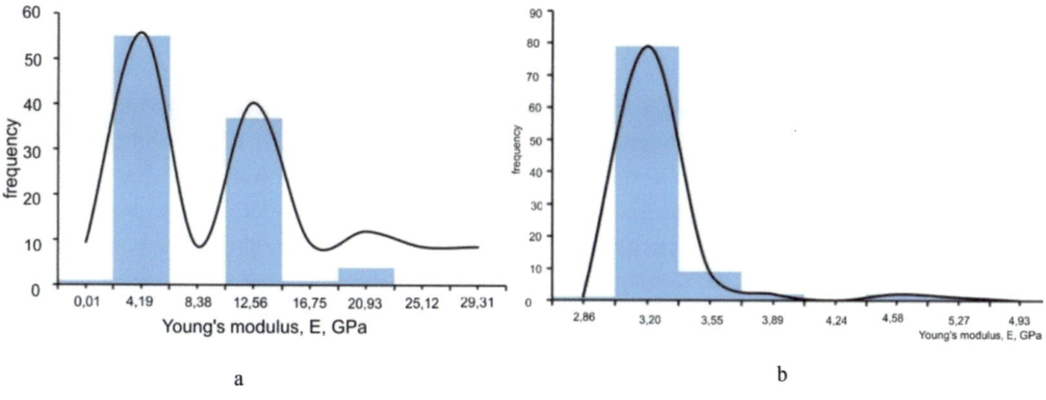

Figure 13. Types of elastic modulus distribution: (a) bimodal for coPUI (R-2300TDI-R)SOD and (b) unimodal for coPUI (R-AltTDI-R)SOD.

So, the dependencies of the modulus of elasticity and microhardness on the insertion depth for synthesized coPUI (R-2300TDI-R)SOD and (R-AltTDI-R)SOD showed that the investigated coPUI differs dramatically in terms of the values of the physical-mechanical characteristics and the type of their distribution. The wide range of values of E is typical for coPUI (R-2300TDI-R)SOD nanocomposites—up to 35 GPa within the surface layers—that may be explained by the presence of a partially ordered phase near the surface, which is extremely sensitive both to the synthesis conditions (the evaporation rate of the solvent) and to surface effects. Contrary, coPUI (R-AltTDI-R)SOD is characterized by greater uniformity of values (up to 4 GPa) and higher microhardness (up to 0.45 GPa). Again, this sensitivity makes it possible to vary the localization of the phase by introduction of carbon nanofillers. The greatest changes in the elastic properties of coPUI (R-2300TDI-R)SOD nanocomposites were detected with the addition of 1 wt.% SWCNTs.

4. Conclusions

The morphology, tribological, and local mechanical properties of new copolyurethane-imides (coPUIs) and nanocomposites based thereof were studied by AFM and NI methods. AFM method in the contact mode of lateral forces revealed the presence of different phases on the surface of synthesized samples. A relationship has been established between the presence of various contrast phases and the chemical structure of the monomers used for synthesis. The ability to control the roughness at the nanoscale for coPUI films on a free surface and on the surface adjacent to substrate is established, using solid substrates of various natures. These characteristics are very important for obtaining materials with improved tribological properties. The AFM and NI methods have shown that the adhesion force and the specific surface energy for synthesized coPUIs increase exponentially with an increase in the normal loading during indentation.

The NI method established the phase separation of coPUI films, which leads to significantly different values of the microhardness on both sides of the copolymer films. The microhardness on the free surface (facing to the air) is much higher than the microhardness of the surface adjacent to the solid support. These results indicate the importance of controlling the microhardness of both film surfaces at nano- and microscales.

Dependencies of microhardness and modulus of elasticity on the depth of indentation have been obtained for initial coPUI (P-2300TDI-P)SOD and coPUI (R-AltTDI-R)SOD and nanocomposites based thereof containing SWCNTs, graphene, and carbon nanofibers. It was found in our case that for each synthesized coPUI, there is only one type of carbon nanomaterials that exerts the greatest influence on their characteristics. Thus, for coPUI (R-AltTDI-R)SOD, it is graphene, the introduction of which leads to an increase in the modulus of elasticity on the free surface up to 4.14 GPa, and for coPUI (R-2300TDI-R)SOD, it is SWCNT, but in this case, the maximum values of the elastic modulus $E = 42.82$ GPa are observed on the surface adjacent to the substrate.

Analysis of the tribological characteristics of nanocomposites based on coPUI (P-2300TDI-P) SOD showed that VGCF exerts the greatest effect on the decrease in C_{fr} and their introduction in amounts about 1 wt.% leads to a decrease in C_{fr} practically by a factor of 2. On the other hand, when VGCF is introduced into the coPUI (R-AltTDI-R)SOD, C_{fr} remains practically unchanged. And, when this coPUI is filled with graphene, an increase in C_{fr} is observed, apparently because of a significant increase in the roughness of the surface of nanocomposite films.

Since the initial coPUI (P-2300TDI-P)SOD and coPUI (R-AltTDI-R)SOD films exhibit radically different values of the friction coefficient, $C_{fr} = 0.362$ and 0.174 for coPUI (P-2300TDI-P) SOD and $C_{fr} = 0.055$ and 0.066 for coPUI (R-AltTDI-R)SOD, the introduction of WS_2 or WSe_2 nanoparticles leads to different effects. Nanocomposites based on coPUI (R-2300TDI-R)SOD with both types of nanoparticles have lower C_{fr} values than the copolymer matrix, while WSe_2 nanoparticles have a greater effect on tribological properties of this coPUI and reduce the C_{fr}

more than twice. Conversely, the initial coPUI (R-AltTDI-R)SOD has a sufficiently low C_{fr}, and the introduction of nanoparticles WS_2 and WSe_2 leads to its almost doubling.

The efficiency of multi-pass scanning tests is shown for investigating the tribological properties of modified copolymer systems. It is found that coPUI (R-AltTDI-R)SOD and nanocomposites based thereof with SWCNT have the best tribological properties, while SWCNTs are uniformly distributed in the material and cause homogeneous structuring at the nanolevel. As a result of studying the tribological properties of coPUI film surface using AFM method with multi-pass scanning, it has been found that effective methods for improving these properties can be either modification by SWCNTs (up to 1 wt.%) or heating to 30°C.

Our results show that synthesized coPUI films and nanocomposites based thereof are very promising materials for tribological applications and their physical-mechanical properties can be controlled both by temperature and by mechanical action.

Acknowledgements

This research was supported by the grants of the Russian Foundation for Basic Research (RFBR) No. 16-53-00178 and the Belarusian Republican Foundation for Fundamental Research (BRFFR) No. F16R-142.

Author details

Tatiana Evgenievna Sukhanova[1,3*], Tatyana A. Kuznetsova[2], Vasilina A. Lapitskaya[2], Tatiana I. Zubar[2], Sergei A. Chizhik[2], Milana E. Vylegzhanina[3], Aleksandr A. Kutin[3], Andrey L. Didenko[3] and Valentin M. Svetlichnyi[3]

*Address all correspondence to: tat_sukhanova@mail.ru

1 Federal State Unitary Enterprise "S.V. Lebedev Institute of Synthetic Rubber", Saint-Petersburg, Russia

2 A.V. Luikov Heat and Mass Transfer Institute NAS Belarus, Minsk, Belarus

3 Institute of Macromolecular Compounds of the Russian Academy of Sciences, Saint-Petersburg, Russia

References

[1] Al-Ajaj I, Kareem A. Synthesis and characterization of polyimide thin films obtained by thermal evaporation and solid state reaction. Materials Science—Poland. 2016; **34**:132-136

[2] Anishchik V, Uglov V, Kuleshov A, Filipp A, Rusalsky D, Astashynskaya M, Samtsov M, Kuznetsova T, Thiery F, Pauleau Y. Electron field emission and surface morphology of a-C and a-C:H thin films. Thin Solid Films. 2005;**482**:248-252

[3] Zavaleyev V, Walkowicz J, Kuznetsova T, Zubar T. The dependence of the structure and mechanical properties of thin ta-C coatings deposited using electromagnetic venetian blind plasma filter on their thickness. Thin Solid Films. 2017;**638**:153-158

[4] Kuznetsova TA, Chizhik SA, Khudoley AL. Deformation structuring of aluminum films upon microindentation. Journal of Surface Investigation. 2014;**8**(6):1275-1285

[5] Kuznetsova TA, Zubar TI, Sudilovskaya KA, Chizhik SA, Didenko AL, Svetlichnyi VM, Sukhanova TE, Vylegzhanina ME. Investigation of the surface of polymer thermoplastic elastomers with detection of various phases in the lateral forces regime. In: XII Intern. Conf Methodological Aspects of Scanning Probe Microscopy; Minsk: Belaruskaya Navuka; 2016. pp. 137-143

[6] Sukhanova TE, Kuznetsova TA, Vylegzhanina ME, Svetlichnyi VM, Zubar TI, Chizhik SA. Possibilities of using probe methods in the diagnostics of nanomodified thermoplastic elastomers. In: XII Intern. Conf. Methodological Aspects of Scanning Probe Microscopy; Minsk: Belaruskaya Navuka; 2016. pp. 8-17

[7] Chizhik SA, Rymuza Z, Chikunov VV, Jarzabek D, Kuznetsova T. Micro- and nanoscale testing of tribomechanical properties of surfaces. In: Jabłoński R, et al., editors. Recent Advances in Mechatronics. Berlin, Heidelberg: Springer; 2007. pp. 541-545

[8] Kuznetsova TA, Zubar TI, Lapitskaya VA, Sudzilouskaya KA, Chizhik SA, Didenko AL, Svetlichnyi VM, Vylegzhanina ME, Kudryavtsev VV, Sukhanova TE. Tribological properties investigation of the thermoplastic elastomers surface with the AFM lateral forces mode. IOP Conference Series: Materials Science and Engineering. 2017;**256**(012022):114-119

[9] Vylegzhanina ME, Kuznetsova TA, Didenko AL, Zubar T, Sudilovskaya KA, Svetlichnyi VM, Kudryavtsev VV, Chizhik SA, Sukhanova TE. Investigation of nanomaterials based on imide-containing thermoplastic elastomers using AFM methods and nanoindentation. In: XXIV Rus. Conf. On Electron Microscopy (RCEM-2016); 2016. pp. 250-251

[10] Bhushan B, Israelachvili JN, Landman U. Nanotribology: Friction, wear and lubrication at the atomic scale. Nature. 1994;**374**:607-616

[11] Diez-Pascual AM, Gomez-Fatou MA, Ania F, Flores A. Nanoindentation in polymer nanocomposites. Progress in Materials Science. 2015;**67**:1-94

[12] Gerkin RM, Hilker BL. Block copolymers: Segmented. In: Buschow KHJ, et al. editors. Encyclopedia of Materials: Science and Technology. 2nd ed. Amsterdam, New York: Elsevier; 2001. pp. 730-732

[13] Yilgör I, Yilgör E, Wilkes GL. Critical parameters in designing segmented polyurethanes and their effect on morphology and properties: A comprehensive review. Polymer. 2015; **58**:A1-A36

[14] Banu P, Radhakrishnan G. Unsaturated poly(ester-imide)s from hydroxyl-terminated polybutadiene, dianhydride and diisocyanate. European Polymer Journal. 2004;**40**: 1887-1894

[15] Banu P, Radhakrishnan G. Thermoplastic poly(esterimide)s derived from anhydride terminated polyester prepolymer and diisocyanate. Journal of Polymer Science, Part A: Polymer Chemistry. 2004;**42**:341-350

[16] Philip Gnanarajan T, Sultan Nsser A, Padmanabha Iyer N, Radhakrishnan G. Synthesis of poly(urethane-imide) using aromatic secondary amine-blocked polyurethane prepolymer. Journal of Polymer Science, Part A: Polymer Chemistry. 2000;**38**:4032-4037

[17] Krijgsman J, Husken D, Gaymans R. Synthesis and properties of thermoplastic elastomers based on PTMO and tetra-amide. Polymer. 2003:7573-7588

[18] Sukhanova TE, Vylegzhanina ME, Didenko AL, Svetlichnyi VM, Gofman IV, Volkov AY, Kutin AA, Yakushev PN, Bershtein VA. Influence of the introduction of carbon nanoparticles on the structure, morphology and properties of nanocomposites based on copolyurethane-imides. In: XXI Intern. Symp. "Nanophysics and Nanoelectronics", Vol. 1; 2017. pp. 331-332

[19] Sukhanova TE, Vylegzhanina ME, Kuznetsova TA, Svetlichnyi VM, Didenko AL, Shiryaeva TI, Kutin AA, Volkov AYa, Kudryavtsev VV, Chizhik SA. Complex diagnostics of morphology and local mechanical characteristics of hybrid materials based on multiblock (segmented) PEI and MQ-resins by AFM methods and nanoindentation. In: 16th Intern. Symp. "Nanophysics and Nanoelectronics", V. 1; 2015. pp. 292-296

[20] Vylegzhanina ME, Kuznetsova TA, Didenko AL, Zubar T, Sudilovskaya KA, Svetlichnyi VM, Kudryavtsev VV, Chizhik SA, Sukhanova TE. Investigation of nanomaterials based on imide-containing thermoplastic elastomers using AFM methods and nanoindentation. In: XXIV Rus. Conf. On Electron Microscopy (RCEM-2016); 2016. pp. 250-251

[21] US Patent 4929358. Polyurethane-imide membranes and their use for the separation of aromatics from non-aromatics; 1990

[22] US Patent 4944880. Polyimide/Aliphatic Polyesters Copolymers; 1994

[23] Didenko A, Yudin V, Smirnova V, Gofman I, Popova E, Elokhovskii V, Svetlichnyi V, Kudryavtsev V. Modification of the thermoplastic polyheteroarylenes with aliphatic polyethers and polyesters: synthesis and dynamic mechanical properties. Materials, Methods & Technologies. 2014;**8**:31-40

[24] Yudin VE, Bugrov AN, Didenko AL, Smirnova VE, Gofman IV, Kononova SV, Kremnev RV, Popova EN, Svetlichnyi VM, Kudryavtsev VV. Composites of multiblock (segmented) aliphatic poly(ester imide) with zirconia nanoparticles: Synthesis, mechanical properties, and pervaporation behavior. Polymer Science Series B. 2015;**56**(6):919-926

[25] Yudin VE, Smirnova VE, Didenko AL, Popova EN, Gofman IV, Zarbuev AV, Svetlichnyi VM, Kudryavtsev VV. Dynamic mechanical analysis of multiblock (segmental) polyesterimides. Russian Journal of Applied Chemistry. 2013;**86**(6):920-927

[26] Bessonov MI, Koton MM, Kudryavtsev VV, Laius LA. Polyimides Thermally Stable Polymers. N.Y: Consultants Bureau; 1987. 318 p

Chapter 6

High-Magnification SEM Micrograph of Siloxanes

Arzu Erol

Additional information is available at the end of the chapter

http://dx.doi.org/10.5772/intechopen.82076

Abstract

SEM is a powerful and efficient microscopy for the analysis of nanomaterials. Although this imaging technique is common and several standard methods exist for chemical analysis, questions remain about the optimal magnification and voltage to be used. The chemical molecules are relatively sensitive to the electron beam. HMDS is as often as possible utilized for surface treatment at the covering of the photosensitive material on the wafer, and trimethylsilanol is created, together with alkali, by hydrolysis of HMDS. The best viewing condition to HMDS and reaction products of organosilicons. The greatest challenges of working with organosilicons molecules are imaging and characterizing features on such a small scale by SEM. The results support the conclusion that, contrary to what is usually recommended, it is best to determine the structure of organosilicon molecules without spectroscopy. It has been a convenient method for the emergence of the structure of HMDS and reaction products. Many micro/nanofabrication technologies have been invented and developed during the past decades. Indeed, some of them have already been widely applied in the cell biology study. In this section, we introduce and emphasize on several prominent technologies, such as soft lithography, electrospinning, nanostructured patterning technologies (including dip pen, e-beam writing, nanoimprint lithography, nanoshaving, and so on), and three-dimensional fabrications. Over the past decade, nanotechnology research has shown exciting evidence that key biological processes (e.g., osteoblast proliferation, osteoblast gene expression, and initial protein adsorption that control such events) can be easily manipulated by modifying the nanotopography of Ti implants. A table is also presented to highlight the pros and cons of different major technologies.

Keywords: siloxanes, organic molecules

1. Introduction: High-magnification SEM micrograph of siloxanes

Scanning electron microscopy (SEM) is a powerful and efficient microscopy for the analysis of nanomaterials. Although this imaging technique is common and several standard methods exist for chemical analysis, questions remain about the optimal magnification and voltage to be used. The chemical molecules are relatively sensitive to the electron beam. HMDS is as often as possible utilized for surface treatment at the covering of the photosensitive material on the wafer, and trimethylsilanol is created, together with alkali, by hydrolysis of HMDS. This study specifically represents the best imaging condition to HMDS and reaction products of organosilicons. The greatest challenges of working with organosilicon molecules are imaging and characterizing features on such a small scale by SEM. The results support the conclusion that, contrary to what is usually recommended, it is best to determine the structure of organosilicon molecules without spectroscopy. It has been a convenient method for the emergence of the structure of HMDS and reaction products.

The process of imaging of molecules is essential for researchers although the scale of small-molecule chemistry has brought with it, challenges. Specifically, with organic compounds having recently been detailed by spectroscopy and diffraction techniques, advances in electron microscopy have made high-resolution images a possibility. The microscopy is capable of displaying numerous small materials, nanostructured structures, and chemical molecules while not sufficiently to illustrate molecular bonds.

Although organic molecular structures are displayed by developing microscopic techniques, many organic molecule structures are yet to be revealed. The examination of a molecule from its fragments is dependent upon the precise manipulation of these molecules. Organic structure characterization is primarily based upon microscopic techniques that are largely established upon electron or X-rays. To obtain an imaging experiment, there exists various techniques such as optical microscopy (OM), transmission electron microscopy (TEM), X-ray microscopy (XRM), scanning electron microscopy (SEM), energy disperse X-ray spectroscopy (EDX), and atomic force microscopy (AFM). There is a wide earthly spectrum, from the morphology level to the atomic level. Each microscope has the ability to display only a restricted number of fields [1, 2]. The appropriate microscope should be selected for the best characterization. The imaging magnitudes of microscopes vary, as TEM, from approximate 1 Å to 10 μm; STEM, from 1 Å to 100 μm; AFM, from 1 nm to 1 mm; SEM, from 1 nm to 1 mm; and OM, from 100 nm to 1 cm. Scanning electron microscopy is an essentially significant technique for image acquisition in the scientific field [3]. It is vital to be able to decipher the structure of nanosized and micro-dimensional materials for characterization of the materials and to interpret the formation and mechanism. Additionally, the imaging of organic molecules and biological specimens poses a serious challenge for researchers.

The applicability of ultraviolet spectroscopy to organosilicon compounds, utilizing mass, Raman, and other spectroscopy techniques [3]. Alkyl-substituted silanes and siloxanes are not ultraviolet absorbent. SEMs are primarily used to analyze the morphology of materials on the samples [4]. The SEM observation scale ranges from microns to nanometers. The decrease of beam damage and charging proves to have an adverse impact on organics. The previous research has shown that microscopy, along with spectrometry, is highly beneficial in identifying organic components [1]. Unfortunately, these methods are destructive to the sample and make conduction of subsequent tests impossible. However, the nondestructive technique of low-voltage SEM enables the identification of organic components.

http://dx.doi.org/10.5772/intechopen.82076

1,1,1,3,3,3-Hexamethyldisilazane (HMDS) is used as a photoresist adhesion promoter in semiconductor applications. It is often possible to locate HMDS residues in SEM images. Silazanes as HMDS are generally moisture sensitive. In addition to appropriate adhesion and photoresist properties, surface dampness, in addition, is a central point. HMDS is well known as a versatile silylation reagent with the ability to block or protect Si—H, N—H, and O—H bonds. HMDS is converted into trimethylsilanol (TMS) and ammonia by the process of hydrolysis in aqueous mediums, as in the equation $((CH_3)_3Si)_2NH + 2H_2O \rightleftharpoons 2(CH_3)_3SiOH + NH_3$ or directly to hexamethyldisiloxane (HMDSO), and ammonia in an aqueous medium, as equation [5]

$$((CH_3)_3Si)_2NH + H_2O \rightleftharpoons ((CH_3)_3Si)_2O + NH_3 \tag{1}$$

According to Donaldson Company, HMDS is able to be hydrolyzed in air to form TMS and ammonia at approximately 25°C, with 45% relative humidity (6–8). Sonnenfeld et al. [6] investigated atmospheric pressure dielectric barrier discharges, with HMDSO admixed into a combination of a noble gas and a molecular gas (O_2 or N_2). The reaction and molecular structure arrangement is illustrated in **Figure 1** [6].

HMDSO can be formed from a minimum number of reaction products, most likely trimethylsilanol, hexamethyldisilane, pentamethyldisiloxane, heptamethyltrisiloxane, and octamethyltrisiloxane.

HMDS, as the silazanes, is similar to siloxanes, with —NH— replacing —O— and acquired in high yield and purity. As dependent on the molecular structure, in the Si—N pillar, $Si(NH)_3$ and $Si(NH)_2$ are realized and the properties of the silazanes that are dependent on the molecular structure such as various functional groups (Si—H, Si—CH_3, Si—CH=CH_2) and degrees of branching [8, 9]. The silazanes, in a process called silylation, react readily with an active hydrogen on any organic chemical (e.g., alcohol, amine, or thiol). For example, hexamethyldisilazane is composed of two silicon atoms fused to the nitrogen atom. 1,1,1,3,3,3-Hexamethyldisilizane (HMDS) is used as a photoresist adhesion promoter in semiconductor applications.

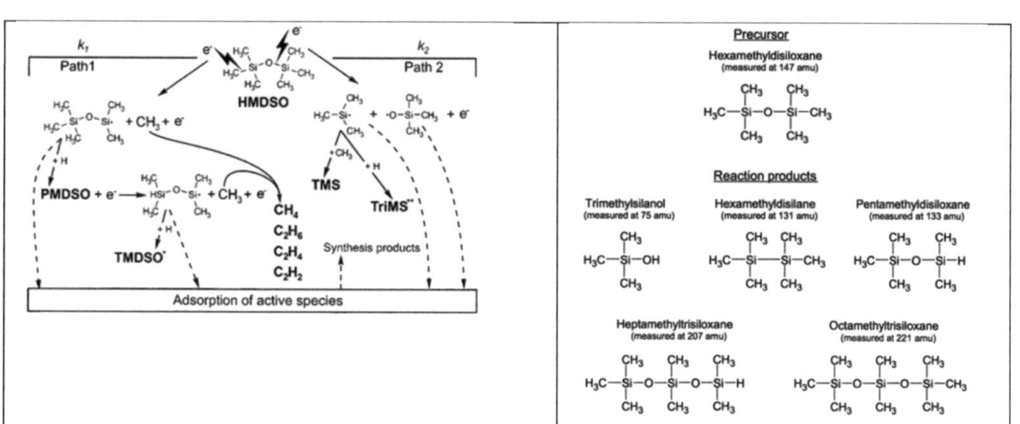

Figure 1. Scheme of the proposed HMDSO reaction chemistry (left) [6]. Possible chemical structures of the identified reaction products in comparison with HMDSO (right) [7].

The Si—O bond is a simple one, in both arrangement and cleavage, an important trait for manufactured science. Weak interactions, along with the simple arrangement and cleavage of Si—O bonded compounds, are used in the construction of supramolecular architectures. The chemistry of the Si—O bond formation is an interesting aspect for hydrolytic cleavage. The Si—O bond can also be compiled from a weak Si—H or Si—Si bond [10–12].

The residue or morphology of small chemical molecules remains unclear, as a result of a lack of high-throughput and high-resolution surface characterization methods. We demonstrate the revealed morphology of surface contaminants by SEM. This study introduces early-stage research about the ability to detect and characterize microscopic fixative residues of HMDS on a biofilm surface, by means of SEM. The results of this research reveal a new possible method of obtaining additional information from the commonly used organic molecules. The results attained from the SEM have proven to be a useful tool in detecting HMDS residues. Advancements in software may increase the abilities of currently available instruments, with microscopy images.

The samples were then dehydrated in a concentration of ethanol, ascending from 50 to 100%, and then dried in hexamethyldisilazane (HMDS, $[(CH_3)3Si]_2NH$, 98.5%; ABCR GmbH & Co. KG, Karlsruhe, Germany) solution overnight. After complete drying, the specimens were sputtered and coated with a thin layer of carbon. The specimen surfaces were examined under a scanning electron microscope XL 30 ESEM FEG, operating at 5 kV and under a magnification rate ranging from 2.000- to 200.000-fold. Tilt angle, spot size, and scanning mode of the electron microscope remained constant for all samples examined. The material was incidentally found when working on the biological surface. HMDS is a drying chemical used for the purpose of fixing scanning electron microscopy specimens.

2. The imaging of surface contaminant, HMDS

Identifying surface residues and contaminants on a biofilm or other biological surface using SEM imaging is seldom explored. One of the major challenges in resolving organic residues, which are the final fixation buffer on a biological surface, is caused by an overlay on top of the image, as both the contaminants and the biological surface are carbon-based materials. In **Figure 2**, HMDS, which is an SEM fixation buffer, appears on the 24 h biofilm formation surface in situ. The topographic contrast exists as a result of the uniformity of the contaminants.

Oral bacteria are visible on the rear surface, and organic residue appears at a lighter concentration on the top surface. The residues therefore prevented the surface from being made visible. The contaminant could be HMDS, and their analogous counterparts as the prepared SEM sample's fixations buffer contain Si.

The water on the biological surface was dried by the evaporation of hexamethyldisilazane. HMDS has been shown in **Figure 3**. During observation of the biological surface, HMDS and residues were measured on dental titanium by means of SEM. Specimens exhibited spontaneous separation during hydrolysis.

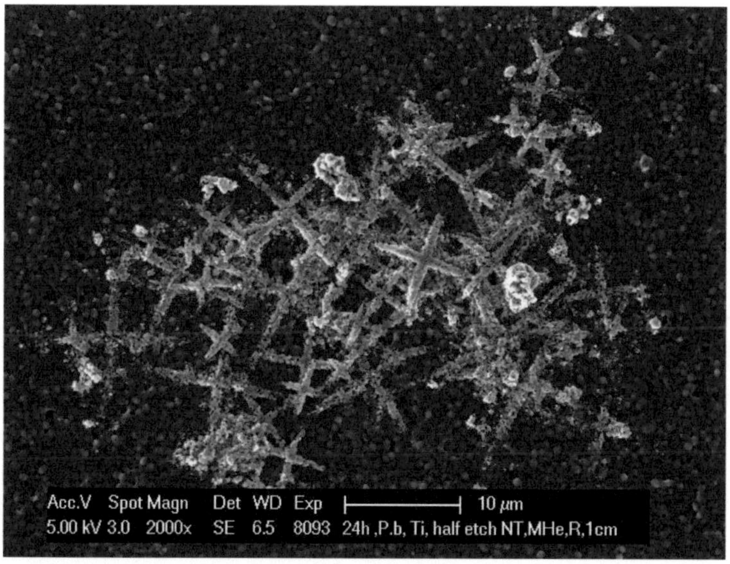

Figure 2. SEM fixative residues on 24 h in situ biofilm formation surface.

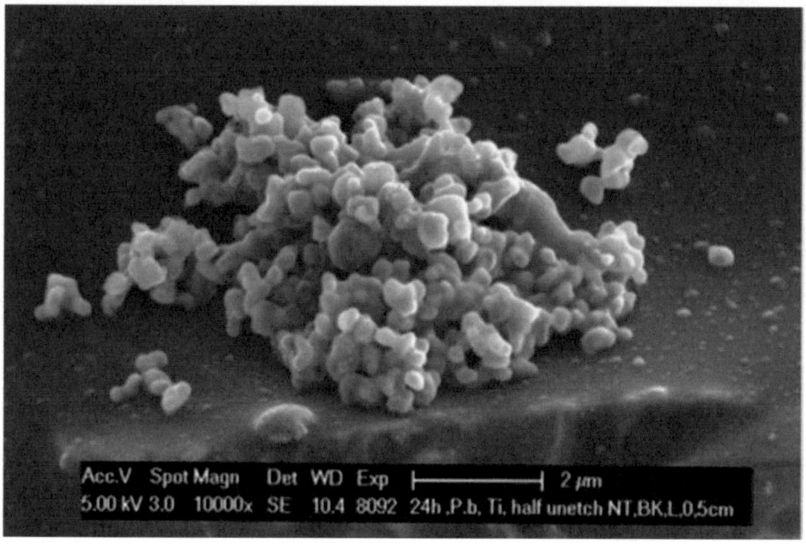

Figure 3. Closer view of the HMDS and related residues on the surface. The shapes of the molecules are seen to be different from each other.

Previous studies had established the same reported morphology for HMDS in this project. The results of this research concur with data obtained by EDX surface analysis, which reflected particle sizes in the range of 20–30 nm [13]. Previous experiments did not show details of HMDS and HMDSO photographed with an electron microscope. As a result of this, it will be a priority for future research and experimentation. Upon reviewing other studies, there is no visual information available about HMDS and HMDSO, although there exists much chemical information about it [14].

3. The imaging of HMDSO transforming into siloxanes

Siloxane molecules are cyclic siloxanes (D_3, D_4, D_5, and D_6) and linear siloxanes (L_2, L_3, L_4, L_5, and L_6). **Table 1** shows that cyclic siloxanes are 2D and 3D conformer generations. In accordance with functional groups, there exist four types of siloxanes. The first of these is the monofunctional (M) units; the second is the difunctional (D) units, which are oligomers' and polymers' linear chains or cyclic compounds; the third is trifunctional (T) units; and the fourth is tetrafunctional (Q) units, which result in branched and spatially cross-linked molecules (elastomers) [2]. The characteristic of these principal structures are presented in **Figure 4**.

Environmental conditions such as temperature, air, and water contact can alter the conformation of siloxanes. The Si—O siloxane chain can be rotated under minimal force at room or higher temperature. When siloxanes are in contact with air, the methyl groups which determine the hydrophobic properties are compacted by the contact surface [16]. When the contact is made with water, the dipole of the siloxane skeleton is responsible for the interactions between the mediums, as the siloxane (silicone) elastomer becomes more hydrophilic [17]. These properties attribute the differences in adhesion properties between silicone elastomers.

Prior studies had proven that silicon does not have the ability to form stable, double bonds. During the synthesis of siloxanes, chains comprising various numbers of repeat units in the chain are formed. Siloxane reaction products [18] are available as oligomers of varying chain length and molecular weights. A single type of molecule does not form at the end of the reaction. The mixture may also contain siloxanes in the cyclic structure. HMDS, the initiator molecule, is converted to siloxanes. As is visible, regular geometric structures vary from HMDS [18] molecules. It is possible that the foliar structures present the linear form of siloxanes molecule [14] and the cyclic form of siloxane molecule, as was shown in **Figure 5**.

Technologies	Pros	Cons	Resolution	Dimension
Soft lithography	Low cost High biocompatibility	Diffusion from the ink can lower the resolution	30 nm–100 mm	2D or 3D
Electrospinning	Suitable for tissue engineering More flexibility in material requirement	Low yield	3 nm–5 µm	3D
Dip pen	Allow the creation of biocompatible nanosized patterns	Limited suitable materials Not suitable for curved surface High cost	30 nm	2D
Electron beam writing	High resolution	Limited suitable materials Not suitable for curved surface High cost	5 nm	2D

Table 1. Summary of the pros and cons of different nanotechnologies [24].

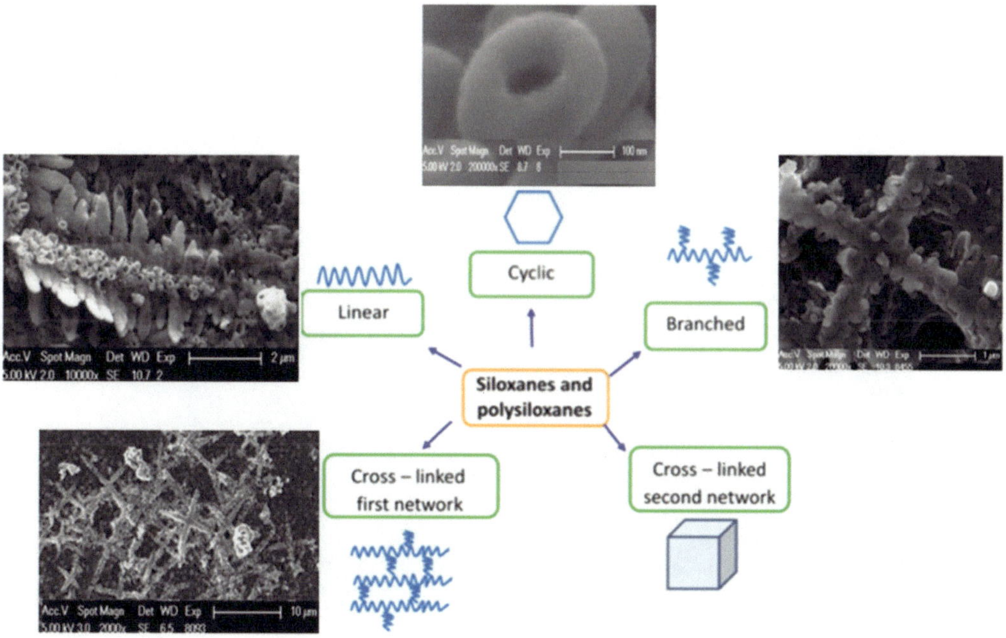

Figure 4. The characteristic and principal structures of siloxanes. The table was drawn inspired from Krystyna M [15].

When closely observing SEM images at the formation of circular structures, it can be seen how the molecule shifts from a linear (lamellar structure) to a circular form. The HMDSO is monitored from the linear structure to the circular structure that was shown in **Figure 6.**

Previous studies have reflected that siloxanes were prepared in many stages and were obtained by means of hydrolysis with excess water at temperatures ranging from 10 to 90°C. For example, continuous hydrolysis of dimethyldichlorosilane (Me_2SiCl_2) produced a mixture of cyclic and linear hydroxyl-terminated oligosiloxanes [15].

The chemical formula of all the circular structures and the electron microscopic image obtained were the groups of molecules that were shown in **Figure 7.**

It is possible to define the measurements of the molecules detected by electron microscopy. To calculate the diameter and volume of the structure, which is estimated to be the group of molecules in the circular form, as is shown in **Figure 8.** The Si—O bond length was 0.164 nm; there were approximately 253 Si—O bonds in the group of molecules.

Synthesis of siloxane elastomers occurs by cross-linking linear siloxane polymers. The process of cross-linking involves the conversion of linear polymers into spatial macromolecules, which are the result of the formation of cross-links, that is, bridges, between them. The poly addition of Si—H bonds to vinyl groups results in the formation of numerous hydrocarbon bridges linking polysiloxane chains. This was the type of cross-linking [15] that was shown in **Figure 9.**

These linear structures lead to the formation of flaccid-appearing structures at the same time. The siloxane chain has an unusually dynamic flexibility; additionally, it possesses a large

Figure 5. SEM image of different types of siloxanes structures.

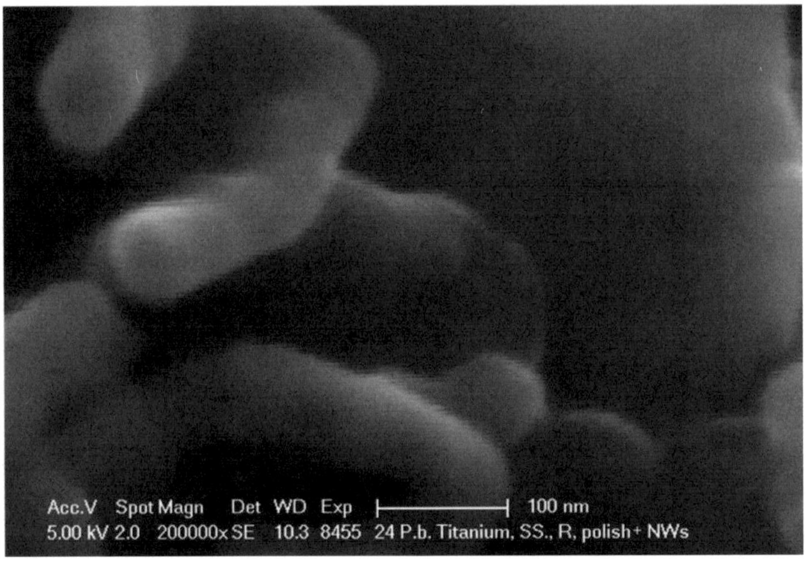

Figure 6. SEM images of HMDSO, from the linear structure to the circular structure. The form could be the groups of molecules that pack together to make these structures.

number of conformations. A conformation can be easily altered. This flexibility facilitates turning by chemical bonds [19]. The conformations may change slightly under stress, leading to the chain conformation adapting to the ambient conditions. Siloxane conformations have minimum free energy of the surface, with these properties possibly creating a thin, highly adherent layer on it (**Figure 10**).

Figure 7. SEM images of the groups of molecules pack together to make these structures.

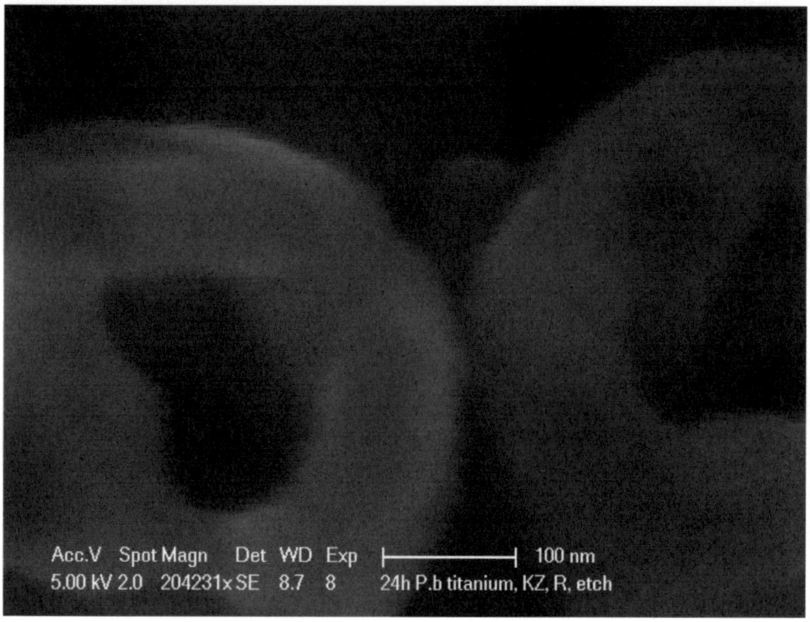

Figure 8. High magnification of SEM images of the groups of molecules packed together to make these structures.

The reaction, beginning with the hydrolysis of the HMDS molecule at the start, is summarized in the reaction table. The transformation of the reaction chain into the HMDSO molecule, of the step-by-step HMDS molecule, is followed by the formation of other siloxane structures from the HMDSO molecule [20]. It is understood that the polymeric structure

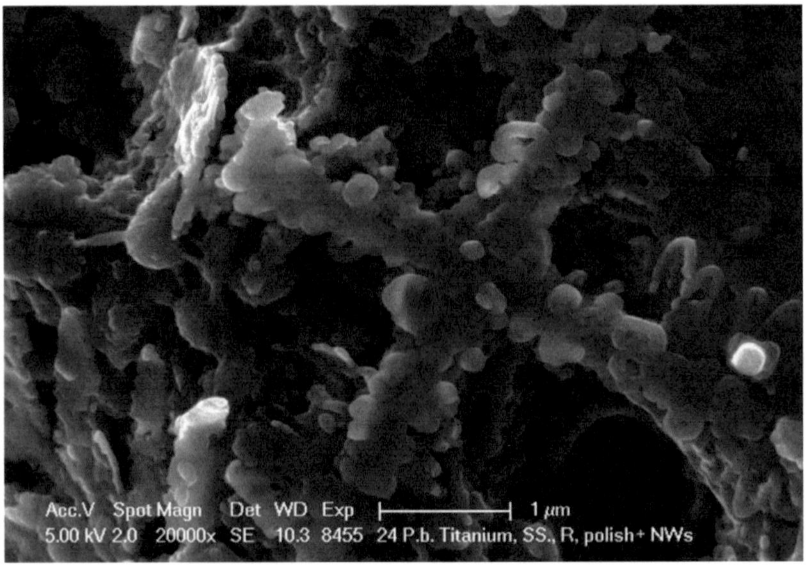

Figure 9. SEM images of the cross-linked linear siloxane polymers.

Figure 10. SEM image of adherent properties of siloxanes on biofilm surface.

has a lamellar appearance, polysiloxanes [21], and a circular structural form. In this work, the images obtained after the coincidence provided a visual display of a chemical reaction. Therefore, it is once again revealed that electron microscopy displays structures with a high magnification power of approximately 100 nm in size.

In this study, during examination of the biological structure on dental implants by electron microscopy, residues were noted on the surface. When considering what this compound may be, it is understood that the electron microscope is the HMDS used in the sample preparation stages and its related properties. Otherwise, the surface containing silicone is free of contaminants.

As a result of closely examining the structures, the states of the HMDS molecule used to provide the fixation on the surface were altered when hydrolysis occurred. It is understood, according to the information in the literature [22], that the molecule resulting from hydrolysis was the HMDSO molecule. Thus, imaging of HMDSO molecules was provided. There exists no similar information in prior studies.

The HMDSO molecule resulting from the hydrolysis of HMDs was observed in various photographs that had been converted to other siloxanes. The structures of the images were predicted from the chemical structure formulas belonging to the silicon structure. It is possible to calculate the number of Si—O bonds [23] in these molecular groups. This study reflected that the reactions of molecules entering the imaging boundaries of electron microscopy can be observed at various stages. Thus, chemical variations can also be visualized by electron microscopy, utilizing simulation studies.

Author details

Arzu Erol

Address all correspondence to: erol.arzu@yahoo.com

Molecular Biology and Genetics, Bulent Ecevit University, Zonguldak, Turkey

References

[1] Jakubikova M, Sadecka J, Kleinova A. On the use of the fluorescence, ultraviolet-visible and near infrared spectroscopy with chemometrics for the discrimination between plum brandies of different varietal origins. Food Chemistry. 2018;**239**:889-897. DOI: 10.1016/j.foodchem.2017.07.008

[2] Neves BR, Salmon ME, Russell PE, Troughton EB Jr. Comparative study of field emission-scanning electron microscopy and atomic force microscopy to assess self-assembled monolayer coverage on any type of substrate. Microscopy and Microanalysis. 1999;**5**(6):413-419

[3] Ponz E, Ladaga JL, Bonetto RD. Measuring surface topography with scanning electron microscopy. I. EZEImage: A program to obtain 3D surface data. Microscopy and Microanalysis. 2006;**12**(2):170-177. DOI: 10.1017/s1431927606060028

[4] Thakkar SV, Allegre KM, Joshi SB, Volkin DB, Middaugh CR. An application of ultraviolet spectroscopy to study interactions in proteins solutions at high concentrations. Journal of Pharmaceutical Sciences. 2012;**101**(9):3051-3061. DOI: 10.1002/jps.23188

[5] Dallas AJ, Ding L, Exley J, Joriman J, Hoang B, Parsons J, et al. Removal of low concentrations of acid gases: Issues and solutions. In: 30th International Microlithography Symposium; 2005. pp. 5752-5119

[6] Sonnenfeld A, Tun TM, Zajíčková L, Kozlov KV, Wagner H-E, Behnke JF, et al. Deposition process based on organosilicon precursors in dielectric barrier discharges at atmospheric pressure—A comparison. Plasmas and Polymers. 2001;**6**:237-266

[7] Reuter R, Reuter K, Ellerweg D, de los Arcos T, von Keudell A, Benedikt J. The role of oxygen and surface reactions in the deposition of silicon oxide like films from HMDSO at atmospheric pressure. Plasma Processes and Polymers. 2011

[8] Hegemann DSU, Fischer A. Macroscopic plasma-chemical approach to plasma polymerization of HMDSO and CH_4. Surface and Coatings Technology. 2005;**200**(1-4):458-462

[9] Theirich D, Soll C, Leu F, Engemann J. Intermediate gas phase precursors during plasma CVD of HMDSO. Vacuum. 2003;**71**(3):349-359

[10] Jal PK, Patel S, Mishra BK. Chemical modification of silica surface by immobilization of functional groups for extractive concentration of metal ions. Talanta. 2004;**62**:1005-1028

[11] Nam NS, Tuan L, Son LT. Synthesis and characterization of organically modified silica nanoparticles by epoxy resin. Journal of Chemistry and Application. 2015;**1**(29):71-74

[12] Zhuravlev L. The surface chemistry of amorphous silica. Colloids and Surfaces. 2000; **173**:1-38

[13] Scimeca M, Bischetti S, Lamsira HK, Bonfiglio R, Bonanno E. Energy dispersive X-ray (EDX) microanalysis: A powerful tool in biomedical research and diagnosis. European Journal of Histochemistry. 2018;**62**(1):2841. DOI: 10.4081/ejh.2018.2841

[14] Yessi J, Ng SC, Osman NAA. Investigation of CPD and HMDS sample preparation techniques for cervical cells in developing computer-aided screening system based on FE-SEM/EDX. Scientific World Journal. 2014:289817. DOI: 10.1155/2014/289817

[15] Mojsiewicz-Pienkowska K, Jamrogiewicz M, Szymkowska K, Krenczkowska D. Direct human contact with siloxanes (silicones)–Safety or risk part 1. Characteristics of siloxanes (silicones). Frontiers in Pharmacology. 2016;**7**:132. DOI: 10.3389/fphar.2016.00132

[16] Soderholm KJ, Shang SW. Molecular orientation of silane at the surface of colloidal silica. Journal of Dental Research. 1993;**72**(6):1050-1054. DOI: 10.1177/00220345930720061001

[17] Flassbeck D, Pfleiderer B, Klemens P, Heumann KG, Eltze E, Hirner AV. Determination of siloxanes, silicon, and platinum in tissues of women with silicone gel-filled implants. Analytical and Bioanalytical Chemistry. 2003;**375**(3):356-362. DOI: 10.1007/s00216-002-1694-z

[18] Meyers VE, Garcia HD, McMullin TS, Tobin JM, James JT. Safe human exposure limits for airborne linear siloxanes during spaceflight. Inhalation Toxicology. 2013;**25**(13): 735-746. DOI: 10.3109/08958378.2013.845629

[19] Clark NM, Garcia-Alvarez P, Kennedy AR, O'Hara CT, Robertson GM. Reactions of (−)-sparteine with alkali metal HMDS complexes: Conventional meets the unconventional. Chemical Communications (Camb). 2009;(39):5835-5837. DOI: 10.1039/b908722b

[20] Jauberteau JL, Jauberteau I. Comparison of hexamethyldisiloxane dissociation processes in plasma. The Journal of Physical Chemistry. 2012;**116**(35):8840-8850. DOI: 10.1021/jp304694z

[21] Enthaler S, Kretschmer R. Low-temperature depolymerization of polysiloxanes with iron catalysis. ChemSusChem. 2014;**7**(7):2030-2036. DOI: 10.1002/cssc.201301386

[22] Hochberg R, Litvaitis MK. Hexamethyldisilazane for scanning electron microscopy of Gastrotricha. Biotechnic & Histochemistry. 2000;**75**(1):41-44

[23] Grabowsky S, Hesse MF, Paulmann C, Luger P, Beckmann J. How to make the ionic Si—O bond more covalent and the Si—O—Si linkage a better acceptor for hydrogen bonding. Inorganic Chemistry. 2009;**48**(10):4384-4393. DOI: 10.1021/ic900074r

[24] Qian T, Wang Y. Micro/nano-fabrication technologies for cell biology. Medical & Biological Engineering & Computing. 2010;**48**(10):1023-1032. DOI: 10.1007/s11517-010-0632-z